Solar Independence: A Comprehensive Guide to Off-Grid Solar Power

Harnessing Sustainable Energy for Off-Grid Living

Olivia Bennett

Table of Contents

INTRODUCTION

Solar Independence stands out as a beacon of hope and innovation in a world increasingly driven by environmental consciousness and a quest for sustainable living. Greetings and welcome to "Solar Independence: A Comprehensive Guide to Off-Grid Solar Power," where we'll explore the revolutionary possibilities of using solar energy to live off the grid.

The need for energy is rising along with the world's population. However, the effects of conventional energy sources on the environment have revolutionized our understanding of and use of power. This e-book is more than simply a guide; it's a manifesto for those who want to live a life unencumbered by conventional electricity systems and are determined to reduce their carbon impact.

The subtitle of this guide, "Harnessing Sustainable Energy for Off-Grid Living," sums it up perfectly. We explore the complexities of solar energy and uncover its potential to free people and communities from reliance on centralized electrical infrastructure. Understanding off-grid life, investigating its benefits, and clearing up common misconceptions are the first steps.

Uncovering the fundamentals of solar energy gives readers a basic understanding of solar photovoltaic systems, how solar panels work, and the variety of solar cell types that are out there. We then guide you through every step of the process of carefully organizing, creating, and setting up your own off-grid solar system, assuring you that it's a feasible and rewarding endeavor. The following chapters include information on the elements essential to the operation of such a system as well as helpful hints for upkeep and leading an off-grid, energy-conserving lifestyle.

We highlight success stories in this book to demonstrate the game-changing potential of solar independence. We examine new developments in off-grid solar power trends as we look to the future and offer a picture of a more ecologically conscious and sustainable world.

This in-depth manual is your road map to attaining solar independence, regardless of your experience with off-grid life or infatuation. Together, let's set out on this adventure to fully realize the promise of sustainable energy for a more promising off-grid future.

CHAPTER I

Understanding Off-Grid Living

Definition and Concept

The idea of "Solar Independence" has evolved as a revolutionary force in the contemporary pursuit of sustainable living, meaning that it is transforming how we generate and consume energy. In its most fundamental sense, solar independence is defined as the capability of producing electricity through solar photovoltaic (PV) systems, thereby freeing individuals and communities from their dependence on conventional power grids. This idea reflects a paradigm shift toward self-sufficiency, which involves utilizing the abundant and renewable energy of the sun to live off the grid.

Exploiting solar photovoltaic systems is the primary characteristic that serves as the defining characteristic of solar independence. Converting sunlight into electricity is called the photovoltaic effect; these systems use solar cells to do this. Solar independence is distinguished from traditional energy sources, dependent on limited fossil fuels, by virtue of this fundamental principle, which reinforces the notion of solar independence. The utilization of solar power, a resource that is not only pure but also renewable and never runs out, is the foundation of this environmentally responsible method of energy consumption.

To achieve solar independence, one of the key goals is to liberate oneself from the limitations and vulnerabilities linked with centralized power grid distribution. Traditional grids are vulnerable to disruptions, which various factors, including natural catastrophes, infrastructure breakdowns, and geopolitical complications, can cause. Individuals can construct robust energy systems that are decentralized and less susceptible to external threats when they embrace solar independence. This not only leads to an increase in energy security but also helps to cultivate a sense of independence and determination.

In addition, the idea of solar independence is ideally in line with broader goals associated with sustainable living and the preservation of the environment. The fact that solar power is a low-impact and low-emission energy source helps reduce the environmental damage caused by traditional energy production. Solar independence is emerging as a feasible way to reduce carbon footprints and contribute to a more sustainable future as societies increasingly struggle to cope with the effects of climate change.

It is important to note that solar independence is not limited to a specific demographic or geographic location. This system may be used in individual households, community developments, and even rural places where access to traditional power infrastructure is limited. It is a versatile and scalable solution. This inclusiveness helps to promote the democratization of energy by making it possible for a wide variety of people and communities to adopt solar independence to their requirements and conditions.

To get a complete understanding of the concept of solar independence, one must first acknowledge the numerous advantages that it offers. Furthermore, in addition to the positive effects on the environment, it also has economic benefits in the long run. Although the initial expenditure for solar photovoltaic (PV) systems is substantial, its long-term savings on energy bills and the possibility of receiving financial incentives from the government make it a financially beneficial decision. Additionally, the increasing economic viability of solar independence results from the growing efficiency of solar technologies and the dropping costs of various solar technologies.

In order to have a complete grasp of solar independence, it is essential to recognize its role in helping to stimulate innovation, notably in the field of energy storage solutions, which is a necessary component of living off the grid. When storing extra energy created during peak solar hours for use during periods of low sunlight or at night, battery technologies, such as lithium-ion batteries, play a crucial role; they are instrumental in this regard. Not only do developments in energy storage make solar independence more reliable, but they also lead to the development of more comprehensive solutions for energy storage that are both efficient and sustainable, which can be applied to a wide range of applications.

Solar independence, on the other hand, is more than just a technological advancement; it is also a shift in mentality toward a more conscientious and responsible attitude toward energy utilization. In recognition of the interconnection of environmental, social, and economic concerns, individuals and communities that adopt solar independence actively craft a sustainable future. As a result, they become advocates for a holistic way of life that goes beyond energy generation and encompasses more significant concepts such as environmental stewardship, resilience, and community empowerment.

As we look more into solar independence, it becomes increasingly clear that it goes beyond technical computation and calculation. It is a manifestation of a vision of a future in which energy is not only a commodity but an essential component of an ecosystem that is both sustainable and linked. The adoption of solar independence represents a purposeful choice to change from being passive energy consumers to becoming active contributors to a cleaner and more resilient energy landscape.

Solar independence also can rethink the socioeconomic dynamics of energy access, which is a significant advantage. The traditional energy infrastructure has frequently contributed to the perpetuation of inequities, with marginalized areas encountering difficulties in gaining access to dependable and affordable power. As a result of the decentralized nature of solar independence, communities can take care of their energy requirements, which helps promote inclusivity and reduce the energy divide.

In conclusion, solar independence is more than just a change in the source of energy; it includes a comprehensive vision for a future environmentally responsible society. It symbolizes a paradigm shift in how we generate and consume energy, fostering self-sufficiency, resilience, and environmental responsibility. Solar independence encompasses a more comprehensive attitude to sustainable living beyond the technical complexities of photovoltaic (PV) or solar energy systems. Individuals and communities that are actively engaged in the adoption of solar independence are not simply installing solar panels; instead, they are contributing to a future energy system that is more robust, equitable, and sustainable. As we negotiate the problems of the 21st century, solar independence emerges as a beacon of hope, presenting a tangible pathway towards a future where energy is harnessed and harmonized with the principles of ecological balance and social equality. This future cannot be imagined without solar independence.

Advantages and Challenges of Off-Grid Living

Off-grid living has become an alternative way of life that disengages individuals and communities from centralized utility networks. This is done in the quest of living in a manner that is both sustainable and self-sufficient. Although this strategy has many potential benefits, it also presents several obstacles that must be considered carefully. By examining the benefits and difficulties associated with off-grid living, one can gain a more nuanced understanding of this way of life, thereby illuminating the transformative potential of this lifestyle and the pragmatic issues accompanying it.

Living off the grid gives individuals greater control over their energy consumption, one of the most significant advantages of this lifestyle choice. Off-grid residents can create electricity by relying on renewable energy sources such as solar panels, wind turbines, or hydropower. This allows them to reduce their reliance on conventional power systems. In addition to fostering a sense of self-sufficiency, this protects persons from power outages and disturbances and ensures that customers have access to a consistent and stable supply of energy.

In addition, living off the grid frequently results in a lower environmental impact. Using renewable energy sources reduces dependency on fossil fuels, which in turn reduces carbon footprints and contributes to the fight against climate change. The deliberate choice to live off the grid is consistent with the concepts of sustainability, which encourage the responsible utilization of resources and the reduction of a negative impact on the ecosystem. Because of their choices in their way of life, those who live off the grid become environmental stewards and serve as a model for living in harmony with surrounding natural elements.

Living off the grid has several significant advantages, including the possibility of cost savings over multiple years. The initial investment in renewable energy infrastructure can be relatively high; however, the continuous running expenses are far lower than those of conventional energy infrastructure. Individuals who live off the grid are freed from the burden of monthly utility payments and the unpredictability of energy costs, which results in a more stable financial outlook for all parties. With time, the return on investment in renewable energy systems becomes apparent, contributing to the stability of the economic situation and the possibility of more lavish spending on discretionary items.

On the other hand, the path to living off the grid is full of obstacles, and the initial expense of constructing a self-sustaining infrastructure is likely one of the most significant of these barriers. A large initial investment is required to install renewable energy sources such as solar panels, wind turbines, and other similar systems. The initial financial strain is further compounded by the expenses associated with purchasing batteries for energy storage and backup generators. Even though these costs can be justified over time through lower utility expenses, the barrier to entry continues to be a significant problem for many people considering living off the grid.

Off-grid life has several challenges, including the difficulty of creating and maintaining an energy system that is dependable and efficient. A certain level of technical competence is required to ensure that the renewable energy system is installed correctly, adequately sized, and considers the local climate and energy requirements. It is essential to do routine maintenance to prevent system failures, and the off-grid inhabitant is responsible for troubleshooting any problems that may arise about the production or storage of energy. To manage the energy infrastructure effectively, committing to continuous education and exercise attention is necessary.

Additionally, living off the grid has its own water and waste management issues. Residents of off-grid communities must build their water supplies, and they frequently rely on wells, rainwater collection, or other local alternatives. This contrasts urban regions, which have centralized water and sewage solutions. Similarly, waste disposal becomes a human duty, necessitating careful thinking to reduce the adverse potential effects on the environment. Regarding these issues, balancing self-sufficiency and responsible resource usage requires careful preparation and dedication to sustainable methods.

Living off the grid has several societal issues, including the fact that it is frequently associated with remoteness and isolation. Living in a location that is further away from urban centers may decrease the number of options for services, healthcare, and education. For individuals who value social interaction and amenities highly, the absence of public infrastructure and community resources can be challenging to overcome. In addition, the psychological component of isolation can affect mental well-being, highlighting the significance of having a strong and resilient attitude for people interested in off-grid life.

Despite these obstacles, living off the grid is not merely a lifestyle choice but a genuine declaration of commitment to more environmentally responsible living. People who live off the grid frequently develop a stronger connection with the natural world, so getting a first-hand awareness of the changes that occur in the environment and the effects that are caused by human activity. One can develop a sense of harmony with the Earth by living off the grid since this lifestyle's daily rhythms align with natural processes.

Living off the grid also encourages a more deliberate and aware approach to consuming, another potential benefit. As a result of restricted resources, individuals need to prioritize their requirements and make intelligent purchasing decisions. This shift toward simplicity and sustainability has broader implications for reducing overall ecological footprints, which contributes to the discourse that is taking place all around the world about living a responsible and ethical standard of life.

When everything is said and done, the benefits and difficulties of living off the grid offer a complicated picture of a way of life beyond merely disengaging from centralized services. People looking for a sustainable and self-sufficient way of life may find that living off the grid is an enticing option because of its autonomy, lessened influence on the environment, and possible cost savings. Nevertheless, there are considerable obstacles to overcome, including the initial expenses, technical complexities, water and waste management difficulties, and social isolation. Off-grid living is not a solution that can be applied universally. Still, for those who choose to start on this road, it is a powerful declaration of dedication to a way of life that prioritizes being environmentally responsible, self-sufficient, and having a profound connection with the natural world. As society continues to struggle with issues of sustainability and resilience, the lessons that may be learned from living off the grid provide vital insights into the possibilities and considerations of a more purposeful and conscious way of life.

Common Misconceptions

While it's true that misconceptions about off-grid living often stem from a lack of understanding, it's important to note the many benefits this lifestyle offers. Living off the grid has gained appeal as it promises sustainability and self-sufficiency, countering the misunderstandings that may skew public perception. This alternative way of life is realistic and deeply satisfying, offering a unique blend of independence and harmony with nature.

Contrary to the common misconception, living off the grid doesn't mean giving up all modern conveniences. Many people who choose this lifestyle continue to use essential technologies and even incorporate renewable energy sources like solar power. The goal isn't to detach from the world but to create a more sustainable, economically independent lifestyle that can be adapted to individual needs and preferences.

An other widespread misunderstanding is the notion that living off the grid is linked with a life of hardship and deprivation. The image of people suffering to satisfy their basic requirements and facing the elements without the conveniences of modern life would discourage some people from considering this way of life. On the other hand, many people who are interested in off-grid living highlight the sense of empowerment and fulfillment that can be achieved by developing a more profound connection with nature and living in harmony with the environment. Living off the grid promotes a mindfulness and conscious approach to consuming, which in turn helps cultivate a sense of appreciation for the fundamental aspects of existence.

In addition, there is a common misunderstanding that living off the grid is a commitment that must be made, requiring individuals to forego the conveniences of metropolitan living entirely. The reality is that life off the grid exists on a scale, with different levels of self-sufficiency throughout the spectrum. Some people may live completely off the grid, while others may adopt off-grid ideas in an urban or suburban environment. Off-grid living is characterized by its adaptability, which enables individuals to personalize their way of life to suit their preferences. This allows them to strike a balance between the benefits of modern life and sustainability.

Living off the grid is not something that is reserved solely for people who are severe survivalists or people who are trying to escape the standards of society, contrary to the widespread notion. Some people choose this way of life to achieve self-sufficiency or react to environmental concerns; nevertheless, others are drawn to it because of the simplicity and peace it provides. People are drawn to off-grid living for various reasons, including the desire to be more thoughtful and deliberate in their lives, be good stewards of the environment, and live a more intentional and mindful existence.

Living off the grid is often misunderstood as unrealistic for those with substantial financial resources. However, with advancements in renewable energy technology, it's more accessible than ever. While costs are associated with setting up off-grid infrastructure, long-term savings on utility bills and potential government incentives can offset these expenses. Moreover, transitioning to off-grid living can be gradual, allowing individuals to adopt sustainable practices without making an immediate, significant financial commitment.

One of the most common misconceptions about living off the grid is that it is connected with a lack of comfort or convenience. In point of fact, a great number of off-grid homes include designs that are both forward-thinking and environmentally friendly, with a focus on both comfort and efficiency. It is possible to live a lifestyle that is not only environmentally friendly but also pleasant and up to date thanks to the development of off-grid technology. These technologies include energy-efficient appliances as well as sophisticated water and waste management systems.

Additionally, there is a common misunderstanding that living off the grid is incompatible with working remotely or having a job. The traditional conception of employment has been revolutionized due to the proliferation of digital connectivity and options for remote work. This has made it possible for individuals who are not connected to the grid to continue their professional commitments. People who live off the grid can strike a balance between their commitment to sustainability and the demands of a modern career if they have access to dependable internet and use technology that allows them to work remotely.

In addition, people tend to misunderstand off-grid life as an escapism activity in which they try to avoid the obligations of being a member of society. In contrast, a significant number of people who are interested in off-grid living are actively involved in their communities and advocate for the principles of sustainability. Those who live off the grid have the potential to serve as a source of motivation for others to embrace more environmentally friendly activities, which can contribute to a more significant societal movement toward environmental consciousness.

In conclusion, it is vital to remove the frequent misunderstandings surrounding off-grid living to cultivate a more accurate understanding of this alternative way of life. The versatility and adaptability of off-grid life allow it to meet various preferences and circumstances despite problems. There is a possibility that individuals will be more likely to investigate this way of life and its chances for living in a manner that is sustainable, self-sufficient, and meaningful if they are aware that living off the grid can be an option that is both realistic and satisfying.

While it's true that misconceptions about off-grid living often stem from a lack of understanding, it's important to note the many benefits this lifestyle offers. Living off the grid has gained appeal as it promises sustainability and self-sufficiency, countering the misunderstandings that may skew public perception. This alternative way of life is realistic and deeply satisfying, offering a unique blend of independence and harmony with nature.

Contrary to the common misconception, living off the grid doesn't mean giving up all modern conveniences. Many people who choose this lifestyle continue to use essential technologies and even incorporate renewable energy sources like solar power. The goal isn't to detach from the world but to create a more sustainable, economically independent lifestyle that can be adapted to individual needs and preferences.

An other widespread misunderstanding is the notion that living off the grid is linked with a life of hardship and deprivation. The image of people suffering to satisfy their basic requirements and facing the elements without the conveniences of modern life would discourage some people from considering this way of life. On the other hand, many people who are interested in off-grid living highlight the sense of empowerment and fulfillment that can be achieved by developing a more profound connection with nature and living in harmony with the environment. Living off the grid promotes a mindfulness and conscious approach to consuming, which in turn helps cultivate a sense of appreciation for the fundamental aspects of existence.

In addition, there is a common misunderstanding that living off the grid is a commitment that must be made, requiring individuals to forego the conveniences of metropolitan living entirely. The reality is that life off the grid exists on a scale, with different levels of self-sufficiency throughout the spectrum. Some people may live completely off the grid, while others may adopt off-grid ideas in an urban or suburban environment. Off-grid living is characterized by its adaptability, which enables individuals to personalize their way of life to suit their preferences. This allows them to strike a balance between the benefits of modern life and sustainability.

Living off the grid is not something that is reserved solely for people who are severe survivalists or people who are trying to escape the standards of society, contrary to the widespread notion. Some people choose this way of life to achieve self-sufficiency or react to environmental concerns; nevertheless, others are drawn to it because of the simplicity and peace it provides. People are drawn to off-grid living for various reasons, including the desire to be more thoughtful and deliberate in their lives, be good stewards of the environment, and live a more intentional and mindful existence.

Living off the grid is often misunderstood as unrealistic for those with substantial financial resources. However, with advancements in renewable energy technology, it's more accessible than ever. While costs are associated with setting up off-grid infrastructure, long-term savings on utility bills and potential government incentives can offset these expenses. Moreover, transitioning to off-grid living can be gradual, allowing individuals to adopt sustainable practices without making an immediate, significant financial commitment.

One of the most common misconceptions about living off the grid is that it is connected with a lack of comfort or convenience. In point of fact, a great number of off-grid homes include designs that are both forward-thinking and environmentally friendly, with a focus on both comfort and efficiency. It is possible to live a lifestyle that is not only environmentally friendly but also pleasant and up to date thanks to the development of off-grid technology. These technologies include energy- efficient appliances as well as sophisticated water and waste management systems.

Additionally, there is a common misunderstanding that living off the grid is incompatible with working remotely or having a job. The traditional conception of employment has been revolutionized due to the proliferation of digital connectivity and options for remote work. This has made it possible for individuals who are not connected to the grid to continue their professional commitments. People who live off the grid can strike a balance between their commitment to sustainability and the demands of a modern career if they have access to dependable internet and use technology that allows them to work remotely.

In addition, people tend to misunderstand off-grid life as an escapism activity in which they try to avoid the obligations of being a member of society. In contrast, a significant number of people who are interested in off-grid living are actively involved in their communities and advocate for the principles of sustainability. Those who live off the grid have the potential to serve as a source of motivation for others to embrace more environmentally friendly activities, which can contribute to a more significant societal movement toward environmental consciousness.

In conclusion, it is vital to remove the frequent misunderstandings surrounding off-grid living to cultivate a more accurate understanding of this alternative way of life. The versatility and adaptability of off-grid life allow it to meet various preferences and circumstances despite problems. There is a possibility that individuals will be more likely to investigate this way of life and its chances for living in a manner that is sustainable, self-sufficient, and meaningful if they are aware that living off the grid can be an option that is both realistic and satisfying.

CHAPTER II

Basics of Solar Energy

Solar Photovoltaic (PV) Systems

Solar photovoltaic (PV) systems stand at the forefront of the global shift towards renewable energy, representing a pivotal technology in pursuing sustainable power generation. An essential component of a solar photovoltaic (PV) system is utilizing the transformational power of sunlight, which is then converted into energy through photovoltaics. This technique is founded on the fundamental idea that certain materials generate an electric current when exposed to sunlight. This principle is the foundation of this process. The technological and environmental significance embedded within these arrays of solar cells can be revealed by understanding the complexities of solar photovoltaic (PV) systems.

The solar cell, sometimes called a photovoltaic cell, is the most critical component of a photovoltaic (PV) system. Solar cells are the fundamental components responsible for capturing sunlight and initiating the electricity generation process. Solar cells are typically constructed from semiconductor materials such as silicon. When photons from the sun reach the surface of the solar cell, they cause electrons to be released from the semiconductor material, which results in the generation of a practical electric current. This flow of electrons makes up the direct current (DC) power generated by the solar cell.

Solar photovoltaic (PV) systems feature inverters, which convert direct current (DC) electricity into a form that may be used in industrial or residential settings. Inverters convert voltage-controlled (DC) power into alternating current (AC), the primary form of electricity utilized in most homes and businesses. This critical conversion guarantees that the electricity generated by the solar photovoltaic system can be integrated without any disruption into the power grids that are already in place or can be used directly to satisfy the requirements for energy.

One of the distinguishing characteristics of solar photovoltaic (PV) systems is how solar cells are organized into modules and panels. These modules, which are more popularly known as solar panels, are made up of several solar cells that are connected and contained within a structure that is both protective and long-lasting. Solar photovoltaic (PV) systems are characterized by their total efficiency and capacity, which are determined by the design and arrangement of solar panels. Solar photovoltaic (PV) systems are becoming increasingly viable and competitive in the energy market due to ongoing increases in efficiency brought about by advancements in solar panel technology.

One of the most significant benefits of solar photovoltaic (PV) systems is their capacity to scale up. Large-scale projects that contribute to the grid might be as modest as off-grid installations for individual residences or as significant as utility-scale projects that contribute to the grid. Because of their scalability, solar photovoltaic (PV) systems can accommodate a wide range of energy requirements, from off-grid locations in remote areas to heavily populated urban centers. Further, the modular design of solar photovoltaic (PV) systems allows them to undergo incremental expansion, enabling customers to expand their capacity per their changing energy needs.

Photovoltaic (PV) solar systems have undeniable positive effects on the environment. Through the utilization of sunshine, a clean and renewable resource, these systems generate power without releasing any greenhouse gases or other pollutants into the atmosphere. As a result of the reduction in carbon emissions, climate change can be mitigated, which is one of the world's most important concerns today. The production, transportation, and installation operations of solar photovoltaic (PV) systems are almost entirely responsible for the carbon footprint that these systems leave behind. As the solar business continues to develop and implement more environmentally friendly practices, the overall impact on the environment is anticipated to diminish, strengthening the green credentials of solar photovoltaic technology.

On the other hand, the widespread implementation of solar photovoltaic systems has its challenges. The sporadic nature of sunlight is one of the most significant obstacles. Solar energy generation depends on the availability of sunshine, which means that solar photovoltaic (PV) systems are most productive during daylight hours. Additionally, the productivity of these systems is influenced by factors such as weather conditions and seasonal variations. Battery technology and other energy storage forms are increasingly included in solar photovoltaic (PV) systems to overcome this difficulty. The extra energy generated during abundant sunlight is stored in these batteries, ensuring a continuous and consistent power supply when there is either a lack of sun or very little sunlight.

Another factor to consider is the capacity of solar photovoltaic (PV) systems. Although there has been a tremendous gain in the efficiency of solar cells due to technological breakthroughs, there is still significant space for improvement. Increasing the amount of sunlight converted into electricity and decreasing the amount of energy lost during the conversion process are necessary steps in improving the efficiency of solar photovoltaic (PV) systems. By increasing the efficiency of solar photovoltaic (PV) systems, which are currently being researched and developed in materials science and engineering, the goal is to make these systems even more competitive with traditional methods of energy generation.

There is a worry regarding land utilization for large-scale solar photovoltaic projects. Significant land tracts are required for utility-scale solar farms, which may affect ecosystems, biodiversity, and agricultural activities. It is of the utmost importance to determine how to meet the requirements for energy while also reducing the adverse effects of solar systems on the surrounding environment. Some examples of innovations that demonstrate efforts to address the land use difficulties connected with solar photovoltaic (PV) systems include floating solar farms and installations that take advantage of bodies of water.

In addition, paying close attention to the environmental and social impacts caused by the production of solar panels and their eventual disposal is highly recommended. The construction of solar panels requires the utilization of a wide range of materials, some of which may have potential adverse effects on human and environmental health. Furthermore, after solar panels end their lives, recycling, and disposal of these panels offer issues that call for environmentally friendly solutions. It is of the utmost importance to address these challenges through appropriate production procedures and efficient recycling programs as the solar industry continues to expand.

Over the years, there has been a substantial improvement in the economic feasibility of solar photovoltaic (PV) systems. This improvement can be attributed, in part, to the reduction in costs associated with manufacture, installation, and continuing maintenance. The government makes the adoption of solar photovoltaic technology more financially feasible through the provision of favorable incentives, subsidies, and favorable regulations. The initial expenditure that is necessary for a solar photovoltaic system continues to be a barrier for some individuals and organizations despite the clear trends that have been observed. To drive wider adoption, overcoming financial obstacles requires a combination of regulatory backing, technology innovation, and public awareness.

Power grids are presented with opportunities and obstacles due to the incorporation of solar photovoltaic (PV) systems into the current energy infrastructure. As a result of the intermittent nature of solar energy output, grid operators are required to handle swings in supply and demand. When it comes to improving the integration of solar photovoltaic (PV) systems into the larger energy landscape, advanced grid management approaches, energy storage solutions, and smart grid technologies all play significant roles. It is necessary for grid management systems to undergo evolution to guarantee stability and dependability as the proportion of solar energy in the total energy mix increases.

In conclusion, photovoltaic (PV) solar systems are a revolutionary force in moving towards sustainable energy worldwide. The fundamental principle of turning sunlight into electricity has led the way for scalable, ecologically friendly, and economically viable solutions to fulfill the world's expanding energy needs. Even though there are obstacles to overcome, like intermittent efficiency, land use, and environmental concerns, continued research and innovation are consistently working to find solutions to these problems. As photovoltaic (PV) technology continues to progress, incorporating this technology into various energy landscapes holds the potential of a future energy system that is cleaner, more sustainable, and more resilient.

How Solar Panels Work

The marvel of solar panels lies in their ability to convert sunlight into electricity through a process known as the photovoltaic effect. It is possible to understand better the complicated workings of solar panels, which sheds light on the scientific prowess behind these devices and their crucial position in the global pursuit of sustainable energy solutions.

The photovoltaic (PV) cell is the principal component of a solar panel. It is a semiconductor device that catches sunlight and converts it into electrical energy. The PV cell is the heart of a solar panel. PV cells, often made of silicon, are designed to take advantage of the characteristics of particular materials that release electrons when exposed to photons from the sun. The occurrence of this phenomenon causes an electric current to begin flowing, which signifies the beginning of the process of energy conversion.

The capability of the semiconductor to generate an electric field is considered the most critical factor in this energy transition. Two layers of semiconductor material are present in a normal photovoltaic (PV) cell. One of these layers is positively charged (P-type), while the other is negatively charged (N-type-type). Electrons in the N-type layer become excited when sunlight strikes the cell because the energy from photons causes them to become excited. This causes the electrons to move over the cell junction and into the P-type layer. An electric field is produced between the layers as a result of this movement, which in turn causes a flow of electrons to occur, which is what makes an electric current.

Metallic conductive plates are strategically positioned on either side of the photovoltaic cell to allow the extraction and exploitation of the electric current that is being generated. The electrons are directed along an external circuit by these plates, which capture and steer the flow of electrons. The flow of electrons through the circuit results in the generation of electrical power that can be utilized. The fact that this electricity is initially provided in the form of direct current (DC), a sort of electrical flow characterized by the movement of charged particles in a single direction, is a crucial point to keep in mind.

It is necessary to add an inverter to the solar panel system to make this electricity compatible with the alternating current (AC) utilized in most residences and places of business. Whether for direct usage in homes and businesses or for seamless integration with existing power grids, the inverter's job is to transform the direct current (DC) electricity generated by the photovoltaic cells into alternating current (AC) electricity. Applying this conversion method guarantees that the electricity produced by the solar panel system is by the standard power supply provided by the utility.

The solar industry has placed a significant emphasis on research and development efforts to improve solar panels' efficiency. Efficiency is the proportion of sunlight that solar panels can convert into electricity. The system's overall efficiency is affected by several factors, including the quality of the semiconductor materials, the design of the PV cells, the angle and orientation of the solar panels in relation to the sun, and so on. Solar panels are becoming more effective and competitive in the energy landscape due to advancements in materials science and engineering, which are responsible for improving efficiency.

To maximize the amount of energy produced, solar panels' orientation and tilt are critical factors. In the northern hemisphere, solar panels are typically installed facing south, whereas in the southern hemisphere, they are installed facing north. This is done to optimize the amount of sunshine received during the day. In addition, the tilt angle of solar panels is modified according to several factors, including geographical location and seasonal variations, to maximize the amount of sunlight captured. Utilizing tracking technologies that monitor the sun's position throughout the day is an additional way to increase energy production.

Even though the fundamental principles behind the operation of solar panels have not changed, there are numerous varieties of solar panel technology. Monocrystalline, polycrystalline, and thin-film solar panels are the most popular commercially available solar panels. Because they are constructed from a single crystal structure, monocrystalline panels are mostly recognized for their high efficiency and elegant appearance. However, polycrystalline panels, built from several crystal structures, are more cost-effective despite having lower efficiency. Although they are comprised of tiny layers of semiconductor material, thin- film solar panels are advantageous in terms of flexibility and ease of integration; nonetheless, they often have lower levels of efficiency.

Rather than being limited to individual rooftop installations, solar panels have a wide range of applications. Large-scale solar farms, made up of enormous arrays of solar panels, substantially contribute to power generation. Through economies of scale, these utility-scale projects can supply power networks with significant quantities of clean energy resources. In addition, solar panels incorporated into building materials, such as solar roof tiles, are a prime example of this technology's adaptability and integration potential in urban settings.

During its functioning, solar panels have a low impact on the surrounding environment, which is one of the outstanding qualities of these panels. Solar energy panels, in contrast to conventional energy sources, which depend on fossil fuels, generate electricity without releasing harmful pollutants or greenhouse gases into the atmosphere. Solar panels are positioned as a sustainable and ecologically friendly energy option because of this attribute, which coincides with the imperative to cut carbon emissions and battle climate change.

Although it is evident that solar panels have a positive impact on the environment while they are in operation, it is vital to consider the entire lifecycle of solar panels, which includes production, transportation, installation, and disposal at the end of their useful lives. The construction of solar panels requires a wide range of materials, some of which may have potential adverse effects on human and environmental health. On the other hand, ongoing efforts within the solar industry are centered on adopting environmentally responsible production techniques and improving the recyclability of solar panels to reduce the environmental footprint that solar panels leave behind.

Throughout the years, there has been a considerable improvement in the economic viability of solar panels. This improvement has been driven by the reduced costs connected with production, installation, and ongoing maintenance responsibilities. The government makes solar panel technology more financially feasible through the provision of financial incentives, subsidies, and favorable laws. There has been a widespread acceptance of solar energy, which has helped to democratize solar energy. This has been made possible by the decreasing cost of solar panels with advantageous economic incentives.

One of the most significant cultural shifts made possible by the widespread use of solar panels is the decentralization of energy-producing capabilities. Instead of relying entirely on centralized power plants, individuals and organizations can create electricity, contributing to a more resilient and distributed energy infrastructure. The reduction of dependency on a small number of centralized sources and the lessening of the impact of power outages are both benefits of this decentralization, which improves energy security.

The integration of solar panels into smart grid technology is causing a transformation in the way energy is managed and delivered. This transformation is occurring as solar panels become more widespread. Smart grids make it possible to monitor and control energy production and consumption in real-time to maximize efficiency and responsiveness to fluctuations in demand. The incorporation of solar panels into smart grids helps to cultivate a dynamic and adaptable energy landscape, which in turn serves to pave the way for power systems that are more environmentally friendly and efficient.

Utilizing the photovoltaic effect to turn sunlight into electricity, solar panels are a marvel of modern engineering. In conclusion, the functioning of solar panels is a marvel of modern engineering. In recent years, there has been a steady progression in the field of materials research, manufacturing techniques, and system design, all of which have contributed to the improving efficiency and cost of solar panel technology. Solar panels, a cornerstone of the worldwide transition to sustainable energy, provide a solution that fulfills the world's expanding energy demands in an environmentally friendly, renewable, and economically viable way. A future energy system that is cleaner, more sustainable, and more resilient can be achieved by the continued integration of solar panels into various energy landscapes.

Types of Solar Cells

There are many varieties of solar cells, which are the essential building blocks of solar panels. Each specific form of solar cell has its own set of qualities and applications. A solid understanding of the many kinds of solar cells is necessary to successfully navigate the varied landscape of solar energy technologies and maximize the integration of these technologies into the global energy grid.

Among the several varieties of solar cells, monocrystalline solar cells are easily distinguished as being among the most effective and commonly utilized. To achieve a better level of purity and uniformity in the material, these cells are constructed from a single crystal structure, often silicon. The manufacturing procedure entails slicing a cylindrical ingot from a single crystal, which results in the production of individual cells that have a distinct black color and look that is sleek. As a result of their high-efficiency levels, monocrystalline solar cells are particularly well-suited for applications that need a restricted amount of area, such as rooftops of residential buildings. While monocrystalline solar cells are more expensive, their appeal can be attributed to their sleek design and high efficiency, contributing to their popularity.

The opposite of monocrystalline solar cells is polycrystalline solar cells, constructed from numerous crystal structures rather than a single crystal. A rough look with a blue color is produced due to the production process, which entails melting raw silicon and casting it into square molds. Polycrystalline solar cells are an exciting option for large-scale solar projects and applications with abundant area because of their cost-effective production. Even though polycrystalline solar cells have a lower efficiency than monocrystalline solar cells, they are available. The crystal boundaries cause polycrystalline cells to have a textured surface. These crystal boundaries scatter sunlight, which can affect the cell's overall efficiency.

A distinct approach to solar technology is represented by thin-film solar cells. Rather than utilizing crystalline silicon, thin-film solar cells are constructed by depositing a variety of semiconductor materials onto a substrate in thin layers during the manufacturing process. The versatility of materials makes such a wide range of uses and production techniques possible. Cadmium telluride (CdTe), copper indium gallium selenide (CIGS), and amorphous silicon (a-Si) are examples of materials that are frequently utilized in thin-film solar cells. There are several benefits associated with thin-film technology, including significantly reduced production costs and the capability to be incorporated into solar panels that are both flexible and lightweight. Despite this, thin-film solar cells often have a lower efficiency when compared to crystalline silicon cells. As a result, they are primarily utilized in large-scale solar farms and applications with many available areas.

These thin-film solar cells made of cadmium telluride (CdTe) have garnered much attention because they are both efficient and cost-effective in converting sunlight into power. CdTe is a compound semiconductor that is capable of efficiently absorbing sunlight, which enables the manufacturing of solar panels that are thin, lightweight, and flexible. Because of this, the CdTe technology is appropriate for large-scale installations, where factors such as cost and convenience of deployment are of the utmost importance. CdTe is a feasible choice in the solar energy landscape, even though concerns about cadmium's influence on the environment have been expressed. However, recent developments in recycling and ethical manufacturing processes attempt to alleviate these concerns and make CdTe something that may be considered.

The thin-film solar cells made of copper indium gallium selenide (CIGS) provide an alternative to the conventional crystalline silicon cells. These cells offer flexibility and efficiency in a wide range of applications. Copper-indium-gallium-selenium (CIGS) cells are a type of cell that may be deposited on flexible substrates. These cells are made entirely of copper, indium, gallium, and selenium. Because of its flexibility, it is possible to create solar panels that are both lightweight and bendable, which expands the range of applications that could be used. Competitive efficiency levels have been demonstrated by CIGS technology, which makes it appropriate for utility-scale solar farms as well as niche markets where flexibility and aesthetics are of utmost importance.

Within thin-film technology, amorphous silicon (a-Si) thin-film solar cells have been a pioneering technological advancement. Since amorphous silicon does not possess the crystalline structure in conventional silicon cells, the manufacturing procedure for amorphous silicon is relatively straightforward. Solar panels made of amorphous silicon are frequently utilized in applications performed on a smaller scale, such as solar calculators and portable chargers, where the flexibility and lightweight features of the material are helpful. Nevertheless, compared to crystalline silicon cells, amorphous silicon cells often demonstrate a lower efficiency, which restricts their employment in applications requiring a significant amount of energy.

A new category of thin-film solar cells is organic photovoltaic (OPV) cells. These cells are characterized by utilizing organic materials, often polymers or tiny organic molecules, as the active semiconductor layer. OPV cells have several distinct benefits, including their adaptability, low weight, and the possibility of manufacturing procedures that are both low-cost and scalable within the industry. Because of the adaptability of organic materials, it is possible to fabricate solar panels that are thin, lightweight, and transparent. This makes organic photovoltaic (OPV) technology viable for situations where conventional solar cells are not feasible. Optical photovoltaic (OPV) cells are currently less efficient than other types of solar cells; nevertheless, continuous research and development efforts are aimed at improving their efficiency and ultimately making them more commercially viable.

Tandem solar cells, also referred to as multi-junction or multi-layer solar cells, are a sophisticated technology that combines many layers of various semiconductor materials to improve the system's overall efficiency. When a tandem solar cell is constructed, each layer is intended to absorb particular areas of the solar spectrum. This allows the cell to make the most of its sunlight. Compared to single-junction solar cells, tandem solar cells can reach better efficiencies because they are constructed by stacking distinct materials with complimentary absorption characteristics. In the context of space applications, where the requirement for maximal energy generation justifies the high cost of solar cells in a constrained space, this technology shows great promise.

Over the past few years, perovskite solar cells have attracted a lot of attention due to their extraordinary efficiency gains and the ease with which they can be manufactured. Perovskite materials, which are called after a naturally occurring mineral with a structure comparable to that of perovskite, can be manufactured by comparatively straightforward procedures. Since perovskite solar cells can be produced at a low cost while also achieving a high-efficiency level, they are an appealing choice for the development of future solar technologies. However, for perovskite solar cells to become a mainstream and sustainable solar technology, it is necessary to address several issues during manufacturing. These challenges include stability, scalability, and influence on the environment.

In conclusion, the dynamic landscape of solar energy research and development is reflected in the wide variety of technologies that are used for solar cells. To meet the requirements of particular applications and the requirements of the market, every form of solar cell comes with its own distinct set of benefits and difficulties. The solar industry constantly evolves, pushing the frontiers of efficiency, cost-effectiveness, and environmental sustainability. This can be seen in areas such as the high efficiency of monocrystalline cells, the flexibility of thin-film technologies, and the promise of future technologies such as perovskite. As solar technology progresses, incorporating various types of solar cells into an all-encompassing energy strategy holds the potential of a cleaner, more sustainable, and more robust energy system.

CHAPTER III

Planning Your Off-Grid Solar System

Assessing Energy Needs

Energy is the lifeblood of modern civilization, powering industries, homes, and the myriad devices that shape our daily lives. Evaluating energy requirements is an important task that calls for a detailed understanding of a wide range of elements, including population expansion, technology improvements, and environmental considerations. Considering the many facets that comprise the process of evaluating energy requirements, this section investigates the approaches, difficulties, and ramifications linked with this essential endeavor.

When it comes to energy assessment, the delicate relationship that exists between population dynamics and energy consumption is the most critical factor. As a result of the rapid growth of the world's population, by the year 2050, we may have reached a population of about 10 billion people. This spike in population offers a tremendous problem when it comes to predicting and meeting the ever-increasing demands for energy. To create efficient energy policies, it is necessary to have a thorough grasp of the patterns of population increase, the tendencies of urbanization, and the developments in socioeconomic conditions.

In addition, technological improvements play a significant part in determining the need for energy. The rapid advancement of technology has ushered in a period of extraordinary innovation, resulting in the introduction of industries and equipment that require a significant amount of energy. The pervasiveness of digitalization, in conjunction with the spread of electric vehicles and intelligent infrastructure, has increased the need for power. Consequently, determining the amount of required energy necessitates a continual analysis of the technical advances and the influence of these trends on consumption patterns.

One of the most important aspects of modern energy evaluations is the shift toward using renewable energy sources. A growing number of people are realizing that it is imperative to move away from fossil fuels and toward cleaner alternatives as the worry around climate change continues to rise. Incorporating renewable energy sources such as solar, wind, hydro, and geothermal into the energy matrix presents several obstacles and opportunities. To develop a sustainable energy landscape, analyzing the economic viability, scalability, and feasibility of renewable energy sources is vital.

Comprehensive energy modeling is one of the critical approaches that is utilized in the process of determining the requirements for energy. The projection of future energy demands within energy models is accomplished using various mathematical techniques, historical data, and scenario studies. These models provide insights into potential energy use patterns by considering multiple factors, including economic growth, technology advancements, and administrative actions. Nevertheless, the accuracy of these models is contingent upon the quality of the data inputs and the capability to predict circumstances that were not anticipated accurately.

Infrastructure development is an essential component of energy assessment, particularly in developing economies experiencing fast urbanization. Services that need a significant amount of energy are in high demand as cities grow and populations congregate in metropolitan regions. A comprehensive grasp of urban planning, transportation systems, and the incorporation of intelligent technologies is required to assess the infrastructure requirements necessary to accommodate this demand. It is also essential for the evaluation to consider the influence of infrastructure projects on the surrounding environment, thereby achieving a delicate equilibrium between development and sustainability.

Within the context of the evaluation process, energy security is of the utmost importance. Nations can be vulnerable if they depend on a single energy source or a limited selection of energy sources collectively. To reduce the risks connected with geopolitical conflicts, disruptions in supply chains, and natural disasters, it is essential to implement policies such as diversifying the energy mix and investing in energy infrastructures that are not susceptible to disruption. When evaluating energy security, it is necessary to assess energy systems' dependability, availability, and resilience to guarantee a consistent and uninterrupted supply throughout the process.

In the context of energy evaluation, the social component involves questions of accessibility, affordability, and equity. Energy poverty continues to be a widespread problem since a sizeable section of the world's population needs access to energy services that are both dependable and in their price range. It is necessary to analyze socioeconomic inequalities to assess energy demands thoroughly. Particular attention should be paid to developing inclusive policies catering to underserved groups' energy requirements. An all-encompassing strategy for evaluating energy use is required to accomplish the complex task of striking a balance between the pursuit of economic progress and social equality.

When it comes to determining the energy requirements, there are many obstacles to overcome, and one of the most pressing ones is the conflict between economic expansion and environmental preservation. Historically, economic growth has been strongly associated with rising energy consumption, frequently driven by sources that produce significant carbon dioxide. It is necessary to find innovative solutions to achieve a balance between supporting economic growth and reducing the environmental impact. Some examples of such solutions include providing financial incentives for environmentally friendly technologies, enforcing severe emission limits, and creating circular economies.

Considering the global character of energy concerns, international cooperation and governance must be implemented. It is impossible to confine energy assessment to the confines of a single nation because of the linked nature of energy systems and the common environmental concerns that require a coordinated approach. Initiatives such as the Paris Agreement call attention to the significance of international collaboration in the fight against climate change and the development of a sustainable future in terms of energy. Because of this, determining the amount of energy required requires the implementation of national programs and a dedication to international collaborations and agreements.

A comprehensive grasp of demographic trends, technical improvements, environmental issues, and social dynamics is required to determine the energy demands of a population. In conclusion, choosing energy needs is a complex and dynamic process. The policies that are made are heavily influenced by the approaches that are used for energy evaluation. These methodologies include modeling and the construction of infrastructure, as well as issues of equity and security. Because of the difficulties connected with balancing economic growth and environmental sustainability, there is an increased demand for creative solutions and international collaboration. As we traverse the energy environment of the future, it will be essential to take a sophisticated and comprehensive approach to assessing energy demands to develop a global energy system that is both resilient and sustainable.

Site Analysis and Location Considerations

In architecture, urban planning, and various development initiatives, the process of site analysis and location considerations is an essential component. Awareness of the natural environment, the local context, and the human factors that define our living environments is a complex dance that requires a comprehensive awareness of all three. This section looks into the facets of site analysis and location considerations. It investigates the approaches, obstacles, and significant ramifications linked with this fundamental part of design and development.

Site analysis is concerned with understanding the physical properties and characteristics surrounding a particular location in its most basic form. A thorough investigation of the terrain, soil composition, climate, and vegetation is required to accomplish this. To ensure that the constructed environment is in harmony with the natural surroundings, the objective is to extract crucial information that can impact design decisions and ensure that the built environment is harmonious. The analysis will likely include topographic surveys, soil tests, and meteorological research to provide a comprehensive grasp of the distinctive characteristics of the location.

An investigation of the site takes into account not only the natural components but also the built environment and the already operational infrastructure. For this purpose, it is necessary to investigate the existing land use, zoning restrictions, and the built environment in the surrounding area. Studying the structures, transportation networks, and utilities currently in the surrounding area is essential to gain valuable insights into how the new construction can be seamlessly integrated with the existing fabric. It is necessary to understand the historical and cultural value of the location to effectively contribute to the preservation of heritage and the development of a relevant story within the built environment.

To determine whether or not a development project will be successful, the socio-economic dynamics of a location are of the utmost importance. It is important to note that the local demography, economic activity, and social trends influence the form and functionality of the proposed development. It is possible to create spaces that are not only visually beautiful but also meet the practical requirements of the people who live there by doing a comprehensive examination of the needs and ambitions of the community. By applying this human-centered approach, the built environment is guaranteed to improve the quality of life for the individuals it serves.

Environmental sustainability is one of the essential aspects that must be taken into account during site investigation. Responsible design necessitates a thorough investigation of the ecological footprint of a project in this day and age, which /is characterized by an increasing number of worries around climate change and environmental damage. Assessing the effects of the development on the ecosystems, water resources, and biodiversity of the surrounding area is a necessary step in this process. Green infrastructure, energy-efficient design, and waste reduction are some of the measures incorporated into sustainable site planning to minimize damaging effects on the environment.

When it comes to a development project, choosing a location that is suitable for the project is a decision that is loaded with severe ramifications. The concerns of location go beyond the physical characteristics of a site and involve a broader range of elements that have the potential to impact the success and durability of a project dramatically. One of the most critical factors is accessibility, and the proximity of a location to transportation hubs, roads, and public transit is essential in assuring connectivity. To develop a well-rounded and livable community, it is equally necessary to ensure that amenities such as schools, hospitals, and recreational places are easily accessible.

One of the most critical aspects of site considerations is the zoning restrictions and land-use policies and regulations. These restrictions govern how land can be utilized and the kinds of structures that can be constructed in a specific location. The observance of zoning restrictions is not only a legal requirement but also guarantees that the development is by the overarching goal of urban or rural planning. Although it is possible to seek variations and exceptions, it is essential to have a comprehensive awareness of the legislative framework to navigate the complicated landscape of land-use planning successfully.

A site's cultural and historical circumstances are critical factors to consider. The design and development process should be informed by the distinctive character of a location, which is formed by the history, traditions, and cultural practices of that geographic location. The preservation of historical landmarks and the incorporation of aspects that are culturally significant into newly constructed buildings both contribute to a sense of continuity and identity throughout the community. The built environment and its community can become disconnected if these components are ignored, which might have negative consequences.

When it comes to the sphere of commercial and industrial building developments, economic concerns become of the utmost importance in terms of location. The proximity of a company to its suppliers, markets, and trained workers can substantially impact the operational efficiency and competitiveness of the company. In addition, decisions regarding the location of a firm or industrial facility are influenced by factors such as the regulatory environment, the overall economic health of a region, and the tax incentives that are available. A comprehensive grasp of the economic dynamics of both the local and regional levels is required to accomplish the difficult task of striking a balance between responsible development and economic viability during development.

The readiness of the infrastructure is an essential component that might determine the success or failure of a development project. When it comes to the success of any project, it is necessary to have access to essential utilities such as water, power, sewage systems, and other similar facilities. Regarding site considerations, one of the most critical aspects is determining the capacity and resilience of the existing infrastructure and planning for any necessary improvements or developments. The absence of sufficient infrastructure might result in difficulties in terms of logistics, a rise in expenses, and even delays in the completion of the project.

Reconciling competing objectives is frequently at the core of the difficulties that arise throughout site study and location decisions. These difficulties encompass a wide range of obstacles. A holistic and interdisciplinary strategy is required to successfully navigate the complex interplay of numerous factors, address the community's demands while adhering to regulatory standards, and balance economic development and environmental sustainability. Furthermore, the dynamic nature of urbanization and global trends adds an extra degree of complexity, making it necessary for the planning and design processes to exercise flexibility and adaptability.

The repercussions of conducting an accurate site study and considering the positioning of components reach far beyond the scope of the current project. Establishing sustainable, resilient, and dynamic communities is facilitated by developments that have been developed with careful planning and strategic placement. A feeling of place is fostered, economic vibrancy is promoted, and the general quality of life for the residents is improved as a result of their presence. On the other hand, poorly planned projects or positioned in a haphazard manner can lead to environmental degradation, social inequity, and economic inefficiencies.

In sum, site analysis and location considerations are fundamental components in the design and development field. To create places that are visually beautiful, useful, sustainable, and responsive to the community's needs, it is vital to have a comprehensive grasp of the natural environment, the built environment, and the human environment. A collaborative and interdisciplinary approach is required to address the inherent complexities and problems of this process. This method should involve community stakeholders, architects, planners, and engineers. The principles of practical site analysis and location considerations will continue to be essential in the process of constructing a built environment that is both sustainable and harmonious as we continue to traverse the ever-changing landscape of urban and rural growth.

Budgeting for Off-Grid Solar Installation

The global quest for sustainable and renewable energy sources has ushered in a new era, placing off-grid solar installations at the forefront of the transition towards cleaner energy. Off-grid solar systems have evolved as a realistic solution, bringing the promise of energy independence and reduced environmental effects. This is in response to the world's issues posed by climate change and the limited nature of traditional energy resources. This section digs into the complexities of budgeting for off-grid solar installations, examining the fundamental components, financial considerations, and the broader implications of accepting solar power in decentralized energy systems. Specifically, the section focuses on the economic consequences of embracing solar power.

The solar photovoltaic (PV) system is the component that is responsible for turning sunlight into electricity. It is the center of any solar project not connected to the grid. The first step in the budgeting process is to thoroughly analyze the energy requirements of the planned application, which could be a remote dwelling, a rural hamlet, or an industrial site not connected to the energy grid. With an understanding of the energy demand, it is possible to determine the suitable size of the photovoltaic (PV) system. This ensures that the system can satisfy the requirements for electricity while also providing for future expansion. Location, the amount of sunlight that the solar panels are exposed to, and the weather patterns all play a significant part in determining the efficiency and effectiveness of the solar panels. These factors also have an impact on the original investment as well as the long-term benefits that the installation provides.

Battery storage is an essential component of off-grid solar systems because it allows for the intermittent nature of solar power generation to be effectively addressed. Energy storage makes it possible to store excess electricity generated during sunny periods so that it can be utilized when there is little or no sunshine. Batteries have to be chosen with great care, considering their capacity, lifespan, and the particular requirements of the system. It is necessary to consider not only the initial cost of the batteries but also the expense of replacing them over time. This is because batteries have a limited lifespan that varies based on the type of battery and how it is used.

When it comes to the operation of off-grid solar systems, inverters and charge controllers are crucial components. The direct current (DC) electricity generated by the solar panels is converted into alternating current (AC) by inverters. AC is the usual form of electricity utilized in homes and businesses. Charge controllers, on the other hand, are responsible for regulating the charging and discharging of the batteries. They prevent the batteries from being overcharged or deeply discharged, each of which has the potential to shorten their lifespan. To ensure the system's dependability and efficiency, it is vital to use high-grade inverters and charge controllers. Additionally, the expenses of these components must be accounted for through the entire budget.

The costs associated with site preparation and installation are frequently underestimated, even though they are essential to the budgeting process. As part of the proper preparation of the site, it is necessary to make sure that the location is suitable for installing solar panels, with minimal shadowing and optimal orientation to optimize the amount of sunshine received. The workforce, equipment, and materials that are necessary to set up the solar array, install the supporting infrastructure, and integrate the entire system are all included in the installation costs. The necessity of committing sufficient finances to this part of the project cannot be overstated. Skimping on the quality of the installation can result in inefficiencies, a shorter lifespan, and increased maintenance expenses.

In addition to the costs of the hardware components,

the financial concerns of off-grid solar installations include continuing maintenance and operational expenses. To preserve the longevity of the system and ensure that it functions at its highest possible level, it is necessary to do routine inspections, cleaning, and occasional repairs. It is essential to have a comprehensive and practical financial plan to prevent disruptions and guarantee the long-term viability of the off-grid solar installation throughout its operating life. The budget should consider both routine maintenance and the possibility of unforeseen expenses.

One of the most significant factors that can influence the

financial sustainability of off-grid solar projects is the presence of government incentives and subsidies. Several nations and regions encourage the adoption of renewable energy solutions through economic incentives, which may include tax credits, subsidies, or feed-in tariffs. To appropriately evaluate the return on investment and determine whether or not the project is feasible, it is essential to incorporate these incentives into the budget analysis. To navigate potential obstacles and avoid unanticipated delays and expenses, it is vital to have a comprehensive grasp of the regulatory landscape, which includes the procedures for obtaining permits and the regulations governing interconnections.

When developing the budget, it is necessary to consider the socioeconomic environment in which the off-grid solar system will be located. In many instances, solar projects not connected to the grid are carried out in rural or distant areas with limited access to electricity. In addition to including provisions for community participation, training, and capacity building, the budget should reflect the socioeconomic conditions of the community that is the focus of the project. It is essential to ensure that the benefits of solar installation go beyond the provision of electricity and include empowerment, education, and improved livelihoods. This will contribute to the overall success and sustainability of the project.

The overarching objective of off-grid solar installations is not limited to providing electricity; instead, it promotes sustainable development and alleviates energy poverty. Therefore, establishing a budget for off-grid solar projects is an investment in the advancement of both society and the environment. In addition to creating economic activity, the benefits include improved health outcomes, expanded educational possibilities, and greater educational chances. When adequately funded and implemented, off-grid solar projects can improve the quality of life in communities, lower emissions of greenhouse gases, and contribute to a more egalitarian and sustainable future.

Although there are many benefits associated with off-grid solar installations, there are always hurdles to be faced, particularly during the early investment period. The initial expenses associated with solar equipment, batteries, and installation can be substantial, which might be a barrier to entrance for specific groups or people. On the other hand, technological developments, in conjunction with the falling prices of solar panels and batteries, are gradually making off-grid solar more affordable. In addition, new financing options are emerging, such as microfinance, community-based funding, and innovative financing models, to address the financial problems and make off-grid solar installations more practical for various stakeholders.

Solar installations not connected to the grid have a notable characteristic that allows them to meet a wide range of energy requirements. These installations range from small-scale home systems to big community or industrial projects. As a result of its modular design, solar photovoltaic (PV) systems can undergo incremental growth in response to changing energy requirements or financial limitations. The flexibility to adapt to changing circumstances is especially beneficial in areas where the energy demand may increase over time or where limited financial resources are available for initial installation. A significant factor contributing to off-grid solar solutions' long-term sustainability and adaptability is the capability to expand the system according to the requirements.

In conclusion, budgeting for off-grid solar systems is a complex and multi-faceted procedure that necessitates an in-depth comprehension of the technical, financial, and social factors involved. Off-grid solar projects provide a way to achieve energy independence, environmental stewardship, and community development when the globe is increasingly turning toward sustainable energy choices. The budgeting process must consider the many components, such as solar panels, battery storage, inverters, site preparation, and ongoing maintenance, while also considering the broader socioeconomic backdrop and the possibility of government incentives. In the quest for a cleaner and more sustainable energy future on a global scale, off-grid solar systems are an appealing alternative because of the potential benefits they offer despite the problems they present.

CHAPTER IV

Components of Off-Grid Solar Systems

Solar Panels

Solar panels, which are sometimes referred to as photovoltaic (PV) panels, are a symbol of the revolution in renewable energy that is currently taking place. They have captured the imagination of scientists, engineers, and environmentalists alike. By utilizing these cutting- edge gadgets, which are made up of solar cells coupled to one another, sunlight has been converted into power with incredible efficiency. Solar panels are becoming increasingly crucial in transforming the global energy landscape. This occurs when the world struggles to cope with the growing difficulties of climate change, the depletion of resources, and the requirement to transition towards sustainable energy sources. Throughout this section, the intricate workings of solar panels are investigated, including their underlying technology, their impact on the environment, the economic concerns involved, and the transformational potential they have for a cleaner and more sustainable future.

Photovoltaics, a topic that has advanced significantly over several decades of study and development, is the science at the heart of solar panels. The solar cell is the fundamental component needed to construct a solar panel. Solar cells are commonly built from semiconductor materials such as silicon. Electrons within the semiconductor material are excited when sunlight, made up of photons, strikes the surface of these cells. This results in the generation of an electric current. The photovoltaic effect is the fundamental concept that underpins converting sunlight into electricity. This phenomenon is also known as the electric conversion process. Numerous solar cell varieties have been developed due to technological advancements. These solar cells include monocrystalline, polycrystalline, and thin-film solar cells, each with its own set of benefits and applications.

The effectiveness of solar panels has been a primary focus of research since the broad adoption of solar panels needs to maximize the amount of energy that can be converted from sunshine. High-efficiency solar panels may create more electricity for each sun unit, making them more economically viable and environmentally friendly. To improve solar panels' efficiency, advancements have been made in the materials used, the manufacturing methods, and the investigation of novel technologies such as tandem solar cells. The development of these innovations is essential to making solar energy a reliable and competitive source within the more extensive energy mix.

Both praise and criticism have been directed at the impact of solar panels on the surrounding environment. Solar panels, on the other hand, can create electricity without releasing any greenhouse gases, which helps reduce the impact of the energy industry on climate change. Reducing carbon dioxide and other air pollutants emissions is a substantial environmental benefit, particularly when compared to the usual method of generating energy through the use of fossil fuels. However, ecological difficulties are associated with producing solar panels and their landfill disposal. Silicon and rare earth metals are two examples of materials and methods that require a significant amount of energy to be utilized in manufacturing solar cells. Furthermore, if the disposal of end-of-life devices is not effectively managed, it might produce electronic waste. To minimize these environmental problems and ensure that solar panels benefit the environment as a whole, the industry needs to strongly emphasize the development of environmentally responsible production processes and recycling technologies.

Economic factors play an equally important role when it comes to the broad use of solar panels. Over the course of history, the initial expenditure required for installing solar panels has been a barrier for many individuals and enterprises. On the other hand, the price of solar panels has dropped dramatically over the last few decades. This is mainly because manufacturing methods have advanced and economies of scale have been achieved. Grid parity, which refers to the situation in which the cost of power generated by solar panels is comparable to that of conventional energy sources, has become a reality in many locations, leading to an increase in the number of people adopting solar energy. The government encourages Individuals and businesses to invest in renewable energy solutions through financial incentives, tax credits, and subsidies. These measures further strengthen the economic feasibility of solar installations.

Solar panels are helpful for a wide variety of applications due to their adaptability, scalability, and versatility. It is possible to design solar panels to fit a wide range of energy requirements, from residential rooftop installations on a modest scale to substantial solar farms that feed into the grid. Solar power systems not connected to the grid and equipped with battery storage can supply electricity in remote places where conventional power infrastructure is unavailable. Powering electronics such as cellphones, computers, and even satellites has become possible thanks to the portability of solar panels, which have found applications in this area. Solar panels are a revolutionary technology that can be integrated into various situations, contributing to the decentralization and democratization of energy generation. This technological adaptability makes solar panels a transformational technology.

A new idea known as solar architecture has emerged due to incorporating solar panels into the built environment. By incorporating photovoltaic technology into building materials in a seamless manner, such as solar roof tiles and solar windows, it is possible to transform buildings into active energy generators. In addition to its practical applications, solar architecture adds to the aesthetic appeal of buildings by simultaneously combining sustainability and design principles. As a symbol of a future in which the difference between energy infrastructure and the built environment becomes increasingly blurry, this integration is evocative of a future in which structures consume energy and actively produce it.

The technological advancements that have been made in solar panels continue to push the limits of what is feasible. The development of new technologies such as solar fabrics, solar paint, and transparent solar cells exemplifies the creative investigation of integrating solar electricity into ordinary things and surfaces. There is the potential for a society in which energy generation is smoothly integrated into the fabric of our everyday lives, and the notion of ubiquitous solar harvesting holds much promise for this world. Implementing these technologies not only broadens the scope of solar power but also challenges the conventional ideas associated with energy infrastructure. They foresee a future in which the transition to renewable energy is a practical requirement and an integrated part of our everyday lives.

It is a transformative characteristic that connects with broader sustainable development aims that solar panels have the potential to revolutionize energy access in poor places due to their ability to change energy availability. Access to dependable power continues to be a severe obstacle in many regions worldwide, mainly rural and outlying areas. Solar panels offer a solution that is both decentralized and scalable, particularly when they are installed in off-grid setups. This improves the quality of life for individuals and communities, stimulates economic activity, promotes education, and provides support for healthcare services. By the vision of sustainable development that does not exclude anyone, democratizing energy access through solar panels is a step in the right direction.

Incorporating solar panels into the existing energy infrastructure has opportunities and obstacles. To overcome the intermittent nature of solar power generation, energy storage continues to be an essential component. The development of new battery technologies, in conjunction with the implementation of novel storage methods, is necessary to guarantee a constant and dependable supply of electricity, particularly when there is much less sunshine. Grid integration and intelligent energy management systems are essential to maintaining a stable and resilient energy grid. These components are necessary for balancing the fluctuations in solar power output.

The conclusion is that solar panels act as catalysts in the continuing transformation that is taking place worldwide toward renewable and sustainable energy sources. Solar panels are positioned to play a vital role in transforming the energy landscape due to their technological proficiency, which enables them to convert sunshine into power, as well as their economic viability and environmental benefits. Even though there are obstacles, continued research and development, in conjunction with a dedication to environmentally responsible manufacturing and recycling processes, are helping to overcome these concerns. Solar panels have the potential to be transformative in more ways than just the generation of electricity; they hold the key to a future that is more sustainable, equitable, and resilient, one in which the power of the sun becomes a cornerstone of our energy paradigm.

Charge Controllers

Charge controllers are sometimes ignored in solar power systems but are essential components. They play a significant role in ensuring that photovoltaic installations are operated in a manner that is both efficient and trustworthy. In addition to regulating the flow of electricity to batteries and avoiding either overcharging or deep draining, these electrical gadgets also function as protectors of the energy generated by solar panels. Having a solid knowledge of the relevance of charge controllers is becoming increasingly important as the globe moves more and more toward solar energy as a clean and sustainable alternative. This section aims to investigate the complex features of charge controllers, including their function in solar power systems, the many types of charge controllers, their significance, and the broader implications that charge controllers have for the dependability and durability of solar installations.

The requirement to turn sunlight into energy and store it so that it may be used when there is little or no sunshine is at the core of solar power systems. Battery technology is often used to fulfill this storage role. Batteries store the excess energy solar panels generate for later use. This energy transfer process is controlled by charge controllers, which serve as the gatekeepers. The solar panels generate direct current (DC) power whenever sunlight strikes them. To guarantee that the batteries are charged to their full potential, the charge controller is responsible for regulating the voltage and current of the electricity. Most importantly, charge controllers safeguard against deep discharge, which can also cause damage to batteries and shorten their lives. Additionally, charge controllers prevent overcharging, which can cause damage to batteries and shorten their lifespan.

PWM controllers, also known as pulse width modulation controllers, and MPPT controllers, also known as maximum power point tracking controllers, are the primary charge controllers. The more conventional and economical choice is to use pulse width modulation (PWM) controllers. They function by fast switching the current from the solar panel on and off, allowing them to regulate the voltage that is transferred to the battery effectively. Although pulse width modulation (PWM) controllers are excellent for smaller solar installations with less complex energy requirements, their efficiency decreases in circumstances in which the output of the solar array does not precisely match the voltage of the battery.

On the other hand, maximum power point tracking (MPPT) controllers are an example of a more advanced and complex technology. By making dynamic adjustments to the solar panels' electrical working point, maximum power point tracking (MPPT) controllers can maximize the efficiency of energy conversion. Because of this adjustment, the solar panels can function at their highest power point, allowing them to extract the maximum energy from the sun. Controllers that use full power point tracking (MPPT) are handy when the temperatures of solar panels or the sunlight conditions are subject to change. This allows for increased energy harvesting. Although MPPT controllers are more expensive than their PWM counterparts, the better efficiency of MPPT controllers typically results in enhanced energy production. As a result, MPPT controllers are the preferred choice for solar installations that are either larger or operate in harsh environmental conditions.

It is impossible to exaggerate the significance of charge controllers in solar power systems' overall design. The primary purpose of these devices is to protect the batteries, which are an essential component of the energy storage component of solar arrays. The phenomenon known as overcharging, in which the batteries are subjected to a current that exceeds what they can store safely, can result in higher temperatures, a shorter lifespan, and, in severe cases, potentially dangerous situations such as venting or leakage. Once the batteries have reached their maximum capacity, charge controllers stop charging, preventing them from being permanently overcharged. Not only does this safeguard the batteries, but it also guarantees the security and durability of the overall solar power system's longevity.

When batteries are reduced to a level that can cause damage to their chemical makeup, a phenomenon known as deep discharging takes place; this phenomenon is the reverse of overcharging. When the voltage of the solar panels dips to a crucial level, charge controllers disconnect the batteries from the panels. This prevents the batteries from being discharged to a significant degree. Taking this preventative measure protects the batteries against damage that cannot be repaired, and their lifespan within the system is increased. Charge controllers, in essence, perform the role of guardians, ensuring that a delicate equilibrium is maintained between the energy harvested from the sun and the energy stored in batteries securely.

The advantages in efficiency that MPPT controllers deliver make them an appealing option, particularly when maximizing energy harvest is of the utmost importance. These controllers perform exceptionally well in environmental conditions in which solar panels are subjected to varying temperatures, shade, or exposure to sunlight that is not uniform. The capacity of maximum power point tracking (MPPT) controllers to make dynamic adjustments to the electrical operating point guarantees that the solar panels will always work at their highest possible efficiency, leading to the greatest possible amount of energy being transferred to the batteries. The enhanced energy yield that MPPT controllers provide can compensate for their higher initial cost in solar projects that are either larger in scale or located in regions that experience unpredictable weather patterns.

Charge controllers contribute to the overall reliability

and stability of solar power systems, in addition to the immediate benefits of protecting batteries and increasing the amount of energy harvested. Charge controllers reduce the amount of wear and tear on batteries by minimizing overcharging and deep draining. This ultimately results in a reduction in the number of times batteries need to be replaced or maintained. The overall sustainability of solar systems is improved due to this, in addition to the fact that it results in cost savings for the owners of installations. The capacity of charge controllers to effectively manage fluctuations in solar output or environmental conditions helps maintain a continuous and steady power supply, which in turn contributes to the long-term viability of solar energy as a mainstream energy source.

Additionally, charge controllers play a part in the

monitoring and diagnostics of the system and their involvement in the management of batteries. Charge controllers of the modern era are outfitted with various features, including LED indicators, digital displays, and communication interfaces, which enable users to monitor the functioning of their solar power systems. The available monitoring tools offer vital insights into energy production, the batteries' status, and the system's overall health. If there are any problems, charge controllers can facilitate the transmission of diagnostic information, which helps in the timely detection and resolution of any issues. Because of this proactive approach to system monitoring, user awareness is increased, maintenance planning is made more accessible, and solar installations can function in an efficient and trouble-free manner.

The importance of charge controllers is increased when solar power systems are installed in areas not connected to the grid or located in distant places. Solar installations sometimes rely heavily on energy storage in situations like these, where access to the electrical grid is limited or nonexistent. These off-grid configurations are equipped with charge controllers, guaranteeing batteries' dependable charging and discharging. This results in a power source that is both continuous and sustainable. The autonomy that charge controllers provide in the management of energy storage enables increased self-sufficiency, making off-grid solar systems a practical and viable choice for towns, enterprises, and individuals in distant areas.

Beyond immediate technical considerations, the work of charge controllers overlaps with broader themes such as energy access, sustainability, and the global transition towards cleaner and renewable energy sources. Charge controllers play an essential role in protecting the environment. Solar power systems equipped with charge controllers provide a decentralized and resilient energy option for areas with either unreliable or nonexistent electricity networks and where power outages occur frequently. This decentralized nature contributes to the democratization of energy by enabling individuals and groups to harness the sun's power without being dependent on centralized utility infrastructure.

Because the world is currently confronted with the imperative of lowering carbon emissions and combating climate change, implementing solar power, in conjunction with technologies that produce effective charge controllers, has emerged as a crucial solution. Solar installations can reduce reliance on fossil fuels drastically. These installations can be installed on rooftops, in utility-scale solar farms, or integrated into the built environment. Charge controllers play a crucial part in boosting the attractiveness of solar energy as a clean, ecological, and economically viable alternative.

They do this by ensuring that solar power systems operate dependably and efficiently.

In conclusion, charge controllers serve as silent sentinels in solar power systems, ensuring solar energy's efficient, safe, and sustainable harnessing. They are responsible for preserving the environment. The dependability, longevity, and performance of solar installations are all improved due to their function in preventing overcharging and deep discharging and allowing system monitoring. The selection of either pulse width modulation (PWM) or maximum power point tracking (MPPT) controllers is contingent upon the particular requirements and circumstances of the solar power system. Both types of controllers play an essential part in facilitating the global transition towards renewable energy. There is a high probability that charge controllers will undergo additional adjustments as technology advances. These refinements will likely improve their efficiency, intelligence, and adaptability to satisfy the growing demand for energy solutions that are both sustainable and resilient.

Batteries

Batteries, those unassuming powerhouses tucked away in our everyday devices, are increasingly emerging as the linchpin of our modern energy ecosystem. Batteries have evolved beyond their traditional use as portable energy sources for electronic devices to become essential components in large-scale energy storage systems. This is because there is a growing emphasis on renewable and sustainable alternative forms of energy worldwide. This section aims to delve into the multifaceted world of batteries, examining their underlying technology, uses across various industries, environmental considerations, and the revolutionary role that batteries play in determining the future of energy storage.

The fundamental concept behind battery technology is the capacity to store electrical energy in a chemical form and then immediately release it when it is required. There are many different kinds of batteries, each designed to meet the requirements of a particular application and performance. This fundamental principle has been utilized in these batteries. The zinc-manganese dioxide reaction is the primary mechanism that underpins the operation of the ubiquitous alkaline batteries, which are frequently seen in gadgets used in the home. These disposable batteries are easy to use and cost-effective, making them an excellent choice for applications that require low power and intermittent power.

On the opposite end of the spectrum are the modern lithium-ion batteries, which are lauded for their high energy density, lightweight design, and the fact that they can be recharged. As a result of their lithium-cobalt oxide or lithium-iron-phosphate chemical compositions, these batteries have become the industry standard for use in portable electronic devices, electric vehicles (EVs), and grid-scale energy storage. As a result of their capacity to store and distribute enormous amounts of energy effectively, they have established themselves as the driving force behind the search for environmentally friendly energy solutions.

Beyond the limitations of consumer electronics, the landscape of battery uses encompasses a far more comprehensive range of applications. Batteries with a high capacity are essential to the operation of electric vehicles, which are a significant factor in moving toward cleaner modes of transportation. The increasing prevalence of lithium-ion batteries, which provide electric vehicles with exceptional ranges and reduce reliance on conventional internal combustion engines, is a significant factor contributing to the move toward electrification in the automobile industry. Not only does this transformation address issues over air pollution and dependence on fossil fuels, but it also brings to light the significant role that batteries play in transforming the transportation sector.

One of the most significant issues linked with solar and wind power is intermittency, which has been addressed by incorporating batteries into the renewable energy sector, which has been a transformative process. The characteristics of renewable energy sources are fundamentally variable, as they are influenced by elements such as the present weather and the time of day. Batteries are the key to managing these changes because they store extra energy generated during times of abundance and then release it when demand is more significant than the corresponding supply. Grid-scale energy storage, made possible by big battery installations, improves the stability and reliability of renewable energy, making it a more viable and competitive alternative to traditional fossil fuels.

Batteries can withstand intermittent energy in locations not connected to the grid or remote. Battery storage is essential for independent power systems, frequently combined with renewable energy sources such as solar or wind power. Batteries are used to store energy at times when there is little or no generation. These systems provide a dependable and long-term power source to communities located in remote areas not connected to centralized power networks. The potential of batteries as agents of social and economic development is shown by the revolutionary influence that batteries have had in providing energy access to previously inaccessible locations.

Research and development activities are pushing the frontiers of battery technology in response to the ongoing spike in demand for energy storage; this need is expected to continue. The conventional liquid electrolytes in solid-state batteries are replaced with solid alternatives, hailed as the next frontier in energy storage. The anticipated benefits are enhanced safety, increased energy density, and longer cycle life. Quantum leaps in energy storage technology, such as developments in sodium-ion and zinc-air batteries, raise the possibility of a future in which batteries are not merely storage devices but rather critical components of an energy grid that is dynamic and responsive.

Even though batteries provide many benefits, there are still obstacles to overcome, and environmental concerns are significant. When batteries are manufactured and disposed of, there are problems around the depletion of resources, pollution, and the accumulation of electronic waste. The extraction of raw materials to create batteries, including lithium, cobalt, and nickel, has several adverse effects on the environment, including the destruction of habitats and the pollution of water sources. In addition, the recycling and disposal of batteries, particularly those that contain hazardous elements, present difficulties in managing waste. As the need for batteries continues to rise, it is becoming increasingly necessary to address the environmental issues that have been raised.

Efforts are currently being made to reduce batteries' negative impact on the environment. These efforts are centered on the development of greener battery technologies, as well as recycling and sustainable sourcing of batteries. We are researching other materials, such as sodium and magnesium, to lessen our dependency on limited resources like lithium and cobalt. In addition, developments in recycling technologies aim to recover valuable elements from spent batteries, thereby reducing the environmental impact connected with the production and disposal of batteries. Ecological stewardship and technological innovation must come together to define a sustainable course for the future of batteries.

Within the context of the larger energy landscape, the economic aspects of battery technology are intricately mixed. The decrease in the prices of lithium-ion batteries, brought about by economies of scale, technological improvements, and better production efficiency, has been a significant factor in the spread of electric vehicles and the expansion of grid-scale energy storage. Batteries have become more accessible for a variety of uses as a result of the dropping cost curve. These applications include providing power to portable electronic devices and supporting renewable energy initiatives. The economics of batteries will continue to impact the trajectory of adopting renewable energy sources and the move towards a more sustainable energy mix as the market for energy storage evolves.

A growing number of people are beginning to embrace the idea of energy storage as a service within the field of renewable energy. Through this model, third-party suppliers offer energy storage solutions on a subscription basis, separating the ownership of batteries from the end-users. Not only does this technique reduce the initial prices for customers, but it also makes it possible to use batteries effectively across a wide range of applications and users. The concept of energy storage as a service has the potential to make access to cutting-edge battery technologies more accessible to the general public and to speed up the incorporation of renewable energy sources into a variety of societal structures.

The rapidly expanding market for electric vehicles is significantly impacting the dynamics of the battery business. In response to the shift toward electrification in the automobile industry, investments have been made in research and development, battery manufacturing capacity, and gigafactories, which are large-scale production facilities dedicated explicitly to battery production. An escalating competition is occurring to produce batteries with longer ranges, shorter charging periods, and lower costs. The effects of this competition extend beyond the automotive sector. Not only does the electrification of transportation result in a reduction in carbon emissions, but it also serves to emphasize the crucial part that batteries play in determining the next generation of mobility.

Energy storage technologies, such as batteries, are one of the most critical components in tackling the issues brought about by the intermittent nature of renewable energy sources. It is now possible to monitor and regulate energy distribution in real time because of the development of smart grids, which integrate sophisticated sensors, communication networks, and control systems. Batteries, essential components of smart grids, allow for storing energy effectively, distributing loads evenly, and stabilizing transmission lines. This synergy between intelligent grids and batteries helps to build an energy infrastructure that is more resilient, adaptable, and sustainable.

Batteries can revolutionize the world of personal energy management, particularly in energy management. Homeowners can store surplus energy generated throughout the day for use during times of high demand or during grid disruptions by utilizing residential energy storage devices, which are frequently linked with rooftop solar panels. At the same time, these systems improve energy self-sufficiency and contribute to load-shifting, reducing the burden placed on the grid during peak hours. Individuals are given the ability to actively participate in the energy transition as a result of the decentralized character of household energy storage, which is in line with the more significant trend toward distributed energy resources.

In conclusion, batteries are more than simply the power source for our electronic devices; they are also the driving force behind transforming our energy landscape. In the process of transitioning towards a more sustainable and resilient energy future, their numerous applications, including enabling energy storage as a service, easing the integration of renewable energy sources, and providing power to electric cars, highlight the essential role they play throughout this process. Battery technology is advancing to meet the concerns of environmental impact and resource depletion through innovative and environmentally responsible approaches.

We will continue to rethink how we produce, store, and use energy. The development and improvement of battery technologies are inextricably linked to moving towards a greener and more sustainable world. As a result, battery technologies have become important agents of change in the intricate web that is our contemporary energy ecology.

Inverters

Within the complex web of energy systems, inverters play a crucial but sometimes unsung role in power production, distribution, and application. These electronic devices are essential for converting direct current (DC) to alternating current (AC), integrating renewable energy sources seamlessly, connecting to the grid, and completing power suitable for various applications. Inverters are complex devices with many uses. This section explores their underlying technology, applications in several fields, their vital role in renewable energy, and their broader implications for the changing energy landscape.

The transformation of electrical energy from one form to another is at the heart of inverter technology. A single-directional electrical flow known as solar panels, wind turbines, and batteries produces direct current. On the other hand, alternating current, in which the direction of electrical flow alternates periodically, powers most appliances, industrial machines, and the electrical grid itself. By bridging this gap and converting DC power into AC power, inverters guarantee that the electricity produced from diverse sources is suitable for the current appliances and infrastructure.

Solar photovoltaic (PV) systems are a prime example of the importance of inverters in applications related to renewable energy. When solar panels are exposed to sunlight, DC electricity is produced. An essential part of these systems are inverters, which transform the DC power generated by the solar panels into AC power for usage in buildings, offices, and the grid. In this case, solar power systems' total performance and financial sustainability are directly impacted by the efficiency and dependability of inverters.

Instead of being a single technology, inverters are available in various forms, each suited to specific uses and technological specifications. Utility-scale solar farms frequently use central inverters to convert DC power from a solar array's panels into AC power at a single location. In contrast, string inverters function at the string level, transforming DC electricity from a network of interconnected solar panels into AC power. Microinverters, a more modern innovation, are put on individual solar panels, turning DC power into AC power at the source. The benefits and trade-offs of each type of inverter influence decisions regarding system design, efficiency, and maintenance.

The expansion of solar power installations has been strongly correlated with the advancement of inverter technology. Traditional inverters were characterized by fixed operating points, which functioned at a single maximum power point. However, more complex solutions were required due to the temperature and sunlight fluctuations. With the advent of Maximum Power Point Tracking (MPPT) technology, the electrical working points of solar panels could be dynamically adjusted by inverters to optimize energy harvest. This was a game-changer. Because MPPT inverters guarantee that the panels perform at their highest capacity regardless of the outside environment, they increase the efficiency of solar power systems.

Beyond solar energy, wind energy is another area where inverters have a significant impact. They are essential in transforming wind turbines' erratic output into consistent, grid-compatible electricity. Wind turbines produce different amounts of electricity in response to variations in wind speed. Utilizing inverters, generated DC power is transformed into AC power at the necessary voltage and frequency, guaranteeing a steady and dependable supply of electricity to the grid. Because they may be used with various renewable energy sources, inverters play a crucial role in the more significant effort to achieve a clean and sustainable energy future.

Inverters play a significant role in energy storage systems by enabling effective battery charging and discharge. Inverters transform DC power from solar panels or wind turbines into AC power for instant use or battery storage in off-grid or hybrid systems. Inverters reverse the process, changing DC power from the batteries into AC power for consumption or grid injection during low- or no-generation times. Here, inverters are conductors that coordinate the smooth transfer of energy between production, storage, and use.

When integrated into smart grids, inverters play a more significant part in the changing energy landscape. Smart grids utilise advanced communication and control technologies to improve electricity delivery's sustainability, efficiency, and dependability. Because they have grid-support features and communication interfaces, inverters help maintain grid stability by offering ancillary services like voltage regulation and reactive power control. The bidirectional flow of power allowed by inverters enhances grid resiliency and enables the integration of dispersed energy resources into the more significant energy infrastructure.

The revolutionary impact of inverters is not confined to technical factors alone; it incorporates economic and societal elements. When it comes to solar energy, the decreasing price of inverters has helped to lower the entire cost of solar systems. To increase efficiency and longevity while cutting costs, inverter makers have concentrated on advancements like modular designs, better thermal management, and new semiconductor materials. This has been crucial in promoting a global shift towards adopting renewable energy sources by making solar energy more affordable and accessible.

Inverters are crucial in decentralizing power generation, consistent with the more significant trend of energy democratization. When combined with inverters, distributed energy resources—such as solar panels and small wind turbines—allow people, organizations, and communities to produce electricity. In addition to improving energy resilience, this decentralization encourages a sense of ownership and involvement in the shift to a more sustainable energy paradigm.

In places with limited access to centralized power grids, such as rural and remote locations, inverters also help to electrify these areas. When combined with inverters, off-grid solar power systems offer a dependable and expandable way to supply electricity to remote areas not connected to traditional infrastructure. These systems take advantage of the flexibility and modularity of inverters to meet various energy requirements, ranging from small enterprises and lighting to powering agricultural equipment and essential appliances. Inverters' contribution to closing the energy gap aligns with international initiatives to combat energy poverty and advance equitable, sustainable development.

One example of how inverters improve grid resilience is the idea of "islanding," which they facilitate. Inverters with anti-islanding characteristics can recognize a grid outage and automatically disengage from the grid. This keeps maintenance staff safe while permitting solar power installations to provide electricity to nearby loads or the grid. The stability and dependability of the power supply are improved overall by inverters' capacity to build microgrids during grid disruptions.

Inverters have many benefits, but they also have drawbacks. Inverter efficiency—which is calculated as the ratio of AC power output to DC power input—is crucial when designing a system. An inverter's efficiency influences a renewable energy system's overall performance and financial sustainability. The main goals of research and development are to increase inverter dependability, minimize energy losses, and increase efficiency. Furthermore, the lifetime costs of renewable energy installations are impacted by the lifespan and upkeep of inverters, highlighting the need for quality standards and technical improvements.

An inverter's lifespan study also highlights environmental considerations. Concerns regarding resource consumption, electronic waste, and possible environmental effects are raised by the manufacture, usage, and disposal of electronic components found in inverters. As a result, the industry is looking into sustainable manufacturing techniques, materials, and recycling initiatives to reduce the ecological impact of inverters. Ensuring unfavorable environmental implications and upholding the advantages of inverters in increasing renewable energy requires the convergence of technology innovation and environmental stewardship.

Inverters are at the center of the world's shift to a more connected and sustainable energy future. Their ability to convert DC power to AC power is essential for integrating renewable energy sources, grid connectivity, and power-generating decentralization. From energy storage and smart grids to solar and wind power systems, invert

Ers act as catalysts for development, enhancing the energy landscape's democratization, efficiency, and resilience. Inverters will be crucial in determining how electricity is generated, distributed, and used in the future as technology develops further, paving the way for a time when clean and sustainable energy sources will be commonplace rather than just a pipe dream.

CHAPTER V

Designing Your Off-Grid Solar System

Sizing Your System

In the dynamic realm of solar energy, sizing a solar system emerges as a critical step in efficiently harnessing the sun's power. Finding the proper specifications for a solar power system is called sizing. These parameters include the maximum number of solar panels, inverters, and energy storage capacity required to fulfill particular energy requirements. In this section, the complexities of sizing a solar system are discussed. These complexities include issues such as energy requirements, site assessments, technology options, and the economic elements that collectively determine the success of a solar installation.

The first and most crucial stage in the process of sizing a solar system is to gain an understanding of the energy requirements of a particular site or application. It is vital to thoroughly analyze the patterns of energy use, regardless of whether the facility in question is a private dwelling, a business establishment, or an industrial complex. A thorough analysis of past energy bills, determining peak consumption periods, and considering any projected shifts in energy demand are all required steps in this process. Accurate data on energy use serves as the foundation for establishing the capacity of the solar power system that is necessary to either compensate for or complement the conventional grid's electricity supply.

Because environmental conditions significantly impact the efficiency of solar panels, site study is an essential component in the process of sizing a solar system. It is important to note that the solar panels' exposure to sunlight is considerably influenced by their geographical position, tilt, and orientation. In addition to this, it is necessary to consider the presence of impediments, shadowing from neighboring structures or plants, and local weather patterns. The solar potential of a location can be evaluated using sophisticated instruments, such as solar irradiance maps and simulation software. This can result in more accurate predictions of the energy generated.

Technological options are available for selecting solar panels and inverter technologies, further enhancing the sizing process. The capacity and efficiency of solar panels differ depending on the type of solar panel, which might be monocrystalline, polycrystalline, or thin-film, as well as the particular properties of the solar panel. Microinverters, string inverters, and maximum power point tracking (MPPT) inverters are some of the available types of inverters. Inverters are essential for converting the direct current (DC) generated by solar panels into alternating current (AC) that may be used or injected into the infrastructure. The solar power system's overall performance and cost-effectiveness are impacted by several technologies, each of which has its own set of advantages and limits.

When determining the size of a solar system, it is essential to balance the initial costs and the benefits that will accrue over time. It is possible to drastically reduce the initial investment required for solar systems by taking advantage of financial incentives, government rebates, and tax credits. At the same time, the price of solar panels has steadily decreased over the years due to innovations in production methods and economies of scale. To conduct a comprehensive economic analysis, the payback period, return on investment (ROI), and total cost of ownership over the system's lifetime are all considered.

Kilowatts (kW) or megawatts (MW) are the standard units of measurement used to define the size of a solar power system. These units represent the system's maximum output capability. Commercial and industrial systems can extend to several hundred or even megawatts, although residential installations typically range from three to ten kilowatts (kW). The process of sizing a solar system entails establishing the optimal capacity in accordance with the project's energy requirements, location characteristics, and financial restrictions.

The intricacies of scaling a system for a home are illustrated by residential solar installations, which are a frequent starting point for many people adopting solar energy. The first step in the procedure is to conduct a thorough energy audit to understand the electricity consumption patterns inside the household. The total energy consumption is affected by various factors, including the number of people living in the house, how they live, and the kinds of appliances they use. The roof's solar potential is evaluated using a site study, which considers the roof's orientation, tilt, and shade. After that, decisions on technology, such as the kind of solar panels and inverters to use, are determined based on several factors, including available space, expenditures, and efficiency.

The idea of "net metering" has an additional impact on the size of home solar power systems. Through net metering, homeowners can obtain credit for exceeding the electricity generated by their solar panels and then transfer them back into the grid. Because of this mechanism, getting an exact measurement of the system is essential to achieve the highest possible return on investment. Oversizing a system may produce more energy than is required, while undersizing a system may result in the system being dependent on the grid for electricity at times of high demand.

Commercial and industrial solar installations present a distinct set of problems and factors to consider regarding sizing. The energy requirements of these establishments are typically more substantial, necessitating the installation of larger solar power systems. The complexity of the process of sizing increases due to the need to carefully assess various parameters, including load profiles, energy-intensive machinery, and peak consumption times. When designing systems capable of satisfying a large amount of the facility's energy requirements, grid connectivity issues become particularly important.

The sizing process takes on a strategic and multidimensional nature for utility-scale solar systems, ranging from several megawatts to gigawatts in capacity. Important considerations include the availability of land, the potential of solar resources, the infrastructure for grid connection, and the approvals required by regulatory agencies. In order to guarantee that large-scale solar projects are technically feasible and environmentally sustainable, the sizing process is guided by comprehensive feasibility studies and environmental effect evaluations.

An additional layer of complexity is added to the equation for sizing due to energy storage. During intense solar production, energy storage systems, which often take the form of batteries, make it possible to collect and use the excess energy generated. Evaluating the desired level of energy independence, the requirement for backup power during grid failures, and the capacity to store and discharge energy at the ideal periods are all considered when determining the size of energy storage. Energy storage for residential installations may concentrate on optimizing self-consumption or providing backup power in the event of power interruptions. Energy storage can contribute to demand management, load balancing, and grid support services in commercial and industrial contexts.

The ever-changing environment of energy rules and regulations is another factor that impacts the configuration of solar power plants. The economic viability of solar systems can be affected by various factors, including feed-in tariffs, incentive programs, and regulatory frameworks. Keeping up with the latest developments in the regulatory landscape is necessary to evaluate the financial consequences of solar projects appropriately and to make judgments regarding system sizing based on reliable information.

Solar system sizing is continuously influenced by technological breakthroughs, with various innovations in materials, design, and efficiency contributing to an overall improvement in performance efficiency. As a result of their ability to collect sunlight from both the front and the back of the panel, bifacial solar panels offer higher energy production and improved efficiency. Higher conversion efficiencies can be achieved by utilizing tandem solar cells, which bring together various materials to catch a broader spectrum of sunlight. These developments affect the process of sizing to increase the amount of energy that can be generated from a given area of solar panels.

The concept of community solar projects introduces a collaborative approach to system scaling. This method allows several persons or entities to enjoy the benefits of a single solar installation. When it comes to these projects, the process of sizing comprises aggregating the energy requirements and preferences of the participants, with the goal of optimizing the solar system's capacity to fulfill the group's requirements together. The issues that individual participants encounter, such as limited roof space, shade, or financial limits, are addressed via community solar initiatives.

To summarize, the process of sizing a solar power system is a complicated and multi-faceted endeavor that requires a careful examination of the energy requirements, the site's characteristics, the technological options available, the economic factors, and the regulatory frameworks. Accurate sizing is essential for optimizing energy production, increasing return on investment, and ensuring the long-term viability of solar systems. This is true whether the applications are residential, commercial, industrial, or utility-scale. When it comes to adequately capturing the sun's power and contributing to a more sustainable and linked energy future, the process of sizing solar systems will continue to be necessary. Technology will continue to progress, and the energy landscape will change.

Choosing the Right Components

The journey towards harnessing solar power is a multifaceted endeavor, and at its core lies the critical task of choosing the right components. The selection of elements that form the backbone of a solar system is intricately entwined with the effectiveness and efficiency of the system. These components include solar panels, inverters, mounting structures, and energy storage systems. This section aims to investigate the complex process of selecting the appropriate components for solar installations. Specifically, the section will explore the most critical factors, technological breakthroughs, and broader implications that determine the success of such systems.

Undoubtedly, the solar panel is the most critical component of any solar power system. The selection of solar panels is of the utmost importance because they are crucial in transforming sunlight into power. There are many different kinds of solar panels in the market, each with distinctive qualities. The use of monocrystalline panels, renowned for their high efficiency, small design, and long-lasting nature, has started increasing in popularity. Although polycrystalline panels are characterized by slightly lower efficiency, they remain a robust solution for many applications. Polycrystalline panels offer an alternative that is more cost-effective at the same time. As a result of their adaptability and the fact that they are suitable for specific situations, thin-film solar panels offer an additional level of diversity in the selection process. Several considerations go into establishing the ideal configuration, including the amount of space available, the limits of the budget, and the efficiency requirements. The selection of the type of solar panel is dependent upon these criteria.

The inverter technology, which is another essential component, is the one that acts as the conductor in a solar system. Inverters are the devices that are accountable for transforming the direct current (DC) produced by solar panels into the alternating current (AC) utilized in residential settings and distributed through the grid distribution system. For example, the decision between string inverters and microinverters is dependent on the preferences of the system designer as well as the current conditions of the site. Microinverters, mounted to individual solar panels, provide better flexibility and optimization than string inverters, which connect many solar panels in series. String inverters are more commonly used in residential home situations. In addition, maximum power point tracking (MPPT) technology has emerged as a game-changer. This technology enables inverters to dynamically modify the electrical working point of solar panels dynamically, maximizing the amount of energy harvested. In addition to the technical criteria, the selection of inverters involves taking into account various other aspects, including the scalability of the system, its reliability, and the ease with which it can be maintained.

A solar installation's efficiency and longevity are directly proportional to the mounting structures, which are frequently disregarded despite their critical role. For example, when deciding between tracking mounting systems and fixed-tilt mounting systems, there are trade-offs between efficiency and the amount of energy captured. Fixed-tilt structures are immovable, cost-effective, and require low maintenance; nevertheless, they function at a fixed angle, limiting their capacity to monitor the sun's passage. On the other hand, tracking systems follow the sun's path throughout the day. This allows them to maximize the energy captured, but it comes with a higher initial cost and requires more maintenance responsibilities. The decision about the mounting of structures depends on site-specific considerations such as the amount of space available, the potential of solar resources, and the limits of the budget.

As a result of the incorporation of energy storage solutions into solar systems, the resilience and dependability of renewable energy installations have been significantly improved. Batteries can store surplus energy generated during peak sunlight hours for use during periods of little or no solar production. Batteries come in various chemistries, including lithium-ion and lead-acid and new technologies like flow batteries. There are several factors to consider while choosing the appropriate energy storage technology, including capacity, cycle life, efficiency, and safety. Lithium-ion batteries have become increasingly popular in various residential, commercial, and utility-scale applications because of their high energy density, rapid charge-discharge capabilities, and extended cycle life. The ever-evolving landscape of energy storage technologies brings new possibilities, each suited to specific use cases and preferences.

Innovative technologies are redefining the landscape of component choices as the solar sector continues to be at the forefront of their development. The development of bifacial solar panels, which can collect sunlight from both the front and the back sides, is an example of an advancement that aims to increase the amount of energy produced. In the continual search for more excellent conversion rates, tandem solar cells, which combine multiple materials to improve efficiency, contribute to the endeavor's success. Solar trackers are an example of an example of an evolution in mounting structures that are meant to maximize energy capture. Solar trackers are automated devices that orient solar panels towards the sun. This technological advancement highlights the dynamic character of the solar business, which constantly evolves to satisfy the needs of efficiency, sustainability, and economic viability. Component selections are continually changing to suit these goals.

The location of the installation heavily influences the selection of components for a solar installation. The choice of solar panels, inverters, and mounting structures is affected by site-specific circumstances such as climate, topography, and local restrictions. For example, solar panels' endurance and temperature coefficient are critical factors to consider in places that experience high temperatures. In a similar vein, regions prone to shading may benefit from the utilization of microinverters or optimizers to reduce the influence that partial shading has on the overall performance of the system. It is possible to ensure that the solar system will perform very well and last a very long time by ensuring that the component choices align with the site-specific conditions.

In selecting the appropriate components for a solar system, selection is not only based on technical considerations; economic factors are also essential to the decision-making process. As a result of breakthroughs in manufacturing methods and economies of scale, the price of solar panels has consistently decreased over the past several years. There is a general decrease in the cost of solar installations, which, when combined with government incentives, tax credits, and financing choices, contributes to the overall affordability of solar equipment. The choice of components affects the return on investment (ROI) and the payback period, which are necessary economic measures. This highlights the need to adopt a holistic approach that balances the initial costs and the long- term advantages.

The selection process for solar components is further refined through intelligent technologies and monitoring systems. Advanced inverters connected with communication interfaces make real-time monitoring of energy output, system performance, and potential problems possible. Smart meters and energy management systems offer customers insights into energy usage patterns, enabling them to make more informed decisions regarding the sizing of components and the optimization of their performance. The approaches driven by data, made possible by these technologies, improve the overall performance of solar installations and their efficiency and dependability.

The process of selecting solar components is more collaborative by implementing community solar initiative programs. Participating in these efforts, numerous individuals make a collective investment in a shared solar installation, allowing them to reap the benefits of the generated energy. In the context of community solar projects, component choices entail aggregating the energy requirements and preferences of members to optimize the system to fulfill the group's requirements as a whole. This collaborative model solves individual members' obstacles, such as limited roof space, shade, or budget limits. It promotes inclusion and shared advantages by addressing these challenges.

Considerations about the environment are increasingly influencing the conversation on selecting solar components. During the life cycle study of solar panels, inverters, and batteries, the environmental impact of these products is evaluated, beginning with the extraction of raw materials and continuing through the manufacture, installation, and disposal stages of their lives. The manufacturing sector is observing an increasing emphasis on environmentally responsible manufacturing processes, recycling programs, and the utilization of products that are beneficial to the environment. The confluence of technological innovation and environmental conscience is an essential component when it comes to ensuring that ethical and sustainable behaviors support solar installation expansion.

A comprehensive awareness of energy requirements, site circumstances, technical improvements, economic considerations, and environmental implications is required to select the appropriate components for solar systems. In conclusion, choosing the right components for solar systems is a multidimensional process. Every element, from solar panels and inverters to mounting structures and energy storage solutions, plays a significant part in determining the level of success a solar installation achieves. The solar business is continuously changing, so the necessity of making informed and strategic component selections is becoming increasingly pronounced. These choices will shape the trajectory of a future energy system that is both sustainable and interconnected.

System Configurations

One of the most critical factors in pursuing a sustainable energy future is solar power's role as a transformative force. The configuration of solar power systems, which refers to how various components are structured to generate, store, and distribute energy, is the most critical factor in the efficient usage of alternative energy sources. There is a wide variety of system configurations, ranging from grid-tied installations that integrate without problems with the current power infrastructure to off-grid options that offer energy autonomy in isolated regions. This section investigates the complexities of system configurations in solar power. Specifically, the research dives into the nuances of grid- tied, off-grid, hybrid, and decentralized setups and the consequences these configurations have on various energy requirements.

Solar power systems connected to the grid are a widespread and interconnected method of utilizing solar energy. To facilitate the flow of electricity in both directions, these topologies are designed to integrate without disrupting the current electrical infrastructure. Solar panels can generate electricity during sunlight, and any excess power is released back into the grid. This typically results in energy credits or financial compensation through net metering. Electricity is pulled from the grid to satisfy the demand for electricity in the event that solar generation is insufficient. Grid-connected systems are chosen because of their ease of use, efficiency, and economic benefits. As a result, they are an appealing option for residential, commercial, and industrial applications.

In residential settings, grid-tied solar systems allow homeowners to generate their own electricity while maintaining a connection to the grid as a dependable backup source. This setup makes it possible to share extra energy with the grid, resulting in lower monthly electricity bills achieved through net metering arrangements. Grid-tied installations are suited for a wide range of property sizes because of their scalability, and they frequently serve as an entry point for individuals interested in solar energy. Grid-tied systems benefit businesses and industries because they help achieve sustainability goals, save operational costs, and, in certain instances, offer extra revenue streams by selling excess energy.

Although they are superior in efficiency and cost-effectiveness, grid-tied systems have their share of difficulties. Dependence on the grid makes a system vulnerable during power outages. Safety rules frequently require these systems to shut down when there is a disruption in the electricity supply from the grid. At the same time, the economic sustainability of grid-tied arrangements is contingent upon legislative frameworks, rules regarding net metering, and the availability of financial incentives. Despite these factors, the widespread adoption of solar power systems connected to the grid highlights these systems' critical role in the continuous transition to a more sustainable energy source.

In contrast to grid-tied installations, off-grid solar power systems are built for autonomy and provide a solution that can be used independently in regions that do not have access to a centralized power grid. Solar panels, energy storage in batteries, charge controllers, and inverters are typically the components that make up these systems. When the sun is shining, the solar panels produce power, stored in batteries for use throughout the night or when a limited amount of solar energy is available. Systems not connected to the grid are commonly used in isolated areas, rural areas, or as emergency backup options when grid connectivity is either prohibitive or impossible.

Individuals and groups can achieve energy independence by utilizing off-grid setups, which eliminate geographical constraints to the availability of electricity. Solar systems that are not connected to the grid can provide a dependable source of electricity for illumination, communication, and essential appliances in geographically isolated regions where the extension of power lines may not be economically possible. Additionally, these arrangements contribute to environmental sustainability by lowering reliance on conventional generators that are powered by gasoline in off-grid settings. This, in turn, reduces the amount of carbon emissions and pollutants produced.

Nevertheless, off-grid solutions provide their own set of difficulties. The system's precise sizing is essential to guarantee a dependable supply of electricity regardless of the weather conditions or the variations that occur throughout the year. The management and storage of energy are significantly influenced by battery technology, which in turn affects the overall efficiency and cost-effectiveness of the system. In off-grid configurations, maintenance and monitoring are also essential factors to consider. This is because the autonomy of these systems necessitates that users actively manage and troubleshoot any problems that may arise. Despite these obstacles, off-grid solar power arrangements demonstrate solar technology's adaptability in meeting a wide variety of energy requirements.

Hybrid solar power systems are a versatile option that maximizes the benefits of both grid-tied and off-grid configurations. These systems combine elements of both grid-tied and off-grid setups. These setups often include integrating solar panels, batteries, inverters, and grid connectivity. This results in a solution that is both comprehensive and versatile, making it suitable for a wide range of applications. Solar panels can generate electricity and concurrently charge batteries when sunlight is abundant. The excess energy can be fed back into the grid. When the amount of solar energy available is minimal, the need can be met by using the energy stored in batteries or grid electricity.

Hybrid systems are designed to meet the demand for energy resilience by combining the dependability of grid-tied setups with the independence of off-grid options. Using energy stored in batteries, homeowners in residential settings can ensure that their power supply remains uninterrupted even in the event of grid failures. Hybrid setups provide organizations and industries with increased control over energy usage, load management, and cost optimization. Hybrid systems are helpful for various applications because of their versatility, allowing them to be used in both on-grid and off-grid areas.

The success of hybrid designs depends on the system's efficient design, the appropriate sizing of the batteries, and the intelligent management of energy. The performance of hybrid configurations can be significantly improved by implementing intelligent technologies, such as modern inverters equipped with grid support functionalities and energy management systems. Energy consumers can achieve a balance between grid connectivity and energy autonomy thanks to the adaptability of hybrid designs, which aligns them with the increasing needs of energy consumers.

A paradigm change in energy distribution is represented by decentralized solar power setups, which challenge the conventional centralized power generation and distribution model. Typically, these designs require the integration of local distribution networks with smaller-scale solar systems dispersed across several different places. The decentralization of solar power contributes to the democratization of energy by making it possible for individuals, communities, and enterprises to have an active role in creating and consuming clean energy.

Solar panels installed on residential rooftops, community solar projects, and distributed energy resources are all examples of decentralized solar arrangements. Residential solar installations are a prime example of the decentralized concept. These installations allow homeowners to generate their electricity using rooftop panels. In contrast, community solar projects incorporate shared solar installations designed to accommodate several participants. This makes it possible for those without rooftops ideal for solar energy to reap the benefits of solar energy. A decentralized and resilient energy landscape is created by utilizing distributed energy resources. These resources include small-scale wind turbines, batteries, and other renewable technologies.

Several causes are driving the emergence of decentralized approaches to solar power systems. Individuals and communities have been given the ability to invest in solar installations on a smaller scale due to the decreasing costs of solar panels, government policies that support solar energy, and various financing alternatives. Enhanced dependability and autonomy of decentralized systems are benefits of developing energy storage technologies such as batteries. To add insult to injury, the decentralization of solar power aligns with the more significant movement toward energy resiliency, sustainability, and community engagement.

Even though decentralized solar power arrangements offer many advantages, there are obstacles to overcome regarding grid interconnection, regulatory frameworks, and system coordination. Within decentralized networks, the intermittent nature of renewable energy sources presents barriers that must be overcome to balance supply and demand. Innovative grid technologies and advanced energy management systems play a significant role in overcoming these problems. These technologies enable the integration of decentralized solar installations smoothly into the more substantial energy infrastructure.

In conclusion, the various configurations of solar power systems constitute a diverse and constantly growing landscape. These configurations are designed to meet the unique energy requirements of individuals, communities, and industries. Grid-tied configurations offer a high level of efficiency and economic benefits while also integrating smoothly with the power infrastructure that is already in place. It is possible to achieve energy autonomy in remote regions by using off-grid solutions, contributing to sustainability, and overcoming geographical constraints. To maximize the benefits of both grid-tied and off-grid arrangements, hybrid installations offer versatility and durability among their many advantages. Individuals and communities are given the ability to actively participate in the transition to clean energy through the use of decentralized solar power arrangements, which pose a challenge to the conventional model of centralized energy delivery. The delicate interaction of various combinations will form the trajectory of a more sustainable and linked energy future. This will occur as the solar industry continues to advance.

CHAPTER VI

Installation Process

Step-by-Step Guide to Installation

Embarking on the journey to harness solar power is an empowering endeavor, and understanding the step-by-step process of solar installation is crucial for a successful transition to clean and sustainable energy. This section aims to explain the complex processes involved in solar installation. It will guide people, businesses, and communities through the process, beginning with the initial planning and site assessment and continuing through the selection of components, installation, and system maintenance.

The first step in installing solar panels is carefully planning and determining the amount of electricity that will be required. By understanding the patterns of electricity usage in the target area, one may lay the groundwork for estimating the right size and capacity of the solar power system. Analyzing previous energy bills, identifying peak consumption periods, and considering projected changes in energy demand due to changes in lifestyle or future expansions are all possible components of residential installations. A comprehensive energy audit may be carried out to determine load profiles, evaluate the current systems' effectiveness, and locate opportunities for energy optimization in the context of businesses and industries.

After determining the amount of energy required, it is necessary to do a comprehensive site survey to maximize the effectiveness of solar panels. Solar panels are subject to substantial sunshine exposure, significantly impacted by their geographical position, orientation, and tilt. To estimate the solar potential of the location, sophisticated instruments such as solar irradiance maps, shading analysis software, and on-site studies would be of great use. To optimize the siting of solar panels and ensure that they have unimpeded access to sunshine throughout the day, factors such as neighboring structures, vegetation, and local weather patterns are taken into careful consideration that are taken into account.

Following the completion of the design and site evaluation phases, the subsequent step entails selecting the appropriate components for the solar power system. Solar panels, inverters, mounting structures, and energy storage systems are all essential components that must be chosen according to the particular requirements of the installation. Whether you go with monocrystalline, polycrystalline, or thin-film solar panels is determined by several criteria, including its efficiency, the amount of available space, and the limits of your budget. Selecting the appropriate inverter technology, whether string inverters or microinverters, is determined by the system design preferences, the location's characteristics, and the scalability requirements. Mounting structures, whether built for tracking or fixed-tilt setups, are determined by a number of criteria, including the amount of available space, the potential of solar resources, and economic considerations. It is necessary to select battery technologies and capacities suitable for the level of energy autonomy and backup power desired if energy storage is going to be a component of the plan.

After the components have been chosen, the installation phase can begin. This phase includes the installation of solar panels, mounting structures, inverters, and energy storage devices. The installation team, which typically consists of highly trained technicians and solar industry experts, works methodically to guarantee every installed component's accurate positioning and connection. Solar panels are fixed securely on rooftops or ground-mounted structures, with adequate consideration given to the ideal orientation and tilt to capture maximum energy. When establishing the electrical circuit between solar panels, inverters, and the electrical load or grid connection, inverters are strategically located to reduce electrical losses, and wire is methodically connected to construct the circuit.

The installation procedure also involves incorporating safety measures to comply with the requirements of the local government and the industry's standards. Several factors contribute to the safety and dependability of the solar power system, including the installation of protective devices, proper grounding, and electrical isolation. During the installation process, comprehensive quality tests and inspections are carried out to guarantee that every component is operating appropriately and that the system is performing to its full potential.

Solar installations connected to the grid require additional procedures to be taken to achieve a seamless connection with the current electrical infrastructure. In specific configurations, inverters are outfitted with grid support features, which make it easier for electricity to flow in both directions. An inverter connected to the grid synchronizes the solar power system with the grid. This ensures that any extra energy can be put back into the grid and that electricity can be drawn from the grid if solar generation is insufficient. To ensure the secure and compliant incorporation of solar electricity into the existing infrastructure, the connection process requires coordination with utility companies and compliance with grid interconnection regulations.

One of the components of the installation process for hybrid and off-grid systems is the installation of energy storage solutions, which are often in the form of batteries. The extra energy created during the hours of high sunshine is stored in the batteries so that it can be used when there is little or no solar production. The installation of batteries requires the configuration of the suitable capacity, the connection of the batteries to the inverter, and the implementation of a charge controller to manage the charging and discharging cycles. There is also the possibility that off-grid systems would contain backup generators to provide an additional layer of energy resilience during extended periods of low illumination.

The commissioning and testing phase comes after the phase of the physical installation, which is completed. The solar power system is put through a battery of tests to guarantee that all of its components are operating appropriately and performing by the requirements determined during the design phase. Performance testing, electrical safety checks, and verifications of the communication system are all essential components of the commissioning process. Once these tests have been completed, the system will be officially connected to the electrical load, the grid, or both, and it will then begin its path of producing clean and renewable energy.

To guarantee the solar power system's long-term performance and durability, executing post-installation monitoring and maintenance and ongoing monitoring activities is essential. Real-time data on energy output, system efficiency, and possible problems can be obtained from monitoring systems integrated into inverters or distinct platforms with their capabilities. In routine inspections, which are usually carried out every year, it is necessary to check for any indications of wear, clean solar panels to maximize the amount of sunlight they absorb, and complete electrical and safety inspections. Suppose there are any faults or deviations in performance. In that case, executing maintenance and repairs as soon as possible is vital to address any issues and maximize the system's operation.

In conclusion, the step-by-step guide to solar installation embodies a comprehensive process that enables individuals, businesses, and communities to move towards clean and sustainable energy. Every stage of the process, beginning with the planning and site assessment stages and continuing through the careful selection of components, installation, and continuous maintenance, is essential in developing solar power systems. For solar technology to be seamlessly integrated into the existing energy landscape, it is necessary to carefully consider various including those that are technological, environmental, and regulatory in nature.-by-step guide to installation offers a road map for navigating the intricacies of solar energy adoption, paving the way for a more sustainable and linked energy future. As the solar industry continues to change, this guide provides a roadmap for installation.

Safety Measures

Adopting solar energy has become a cornerstone of the global push toward sustainability, offering a clean and renewable alternative to conventional power sources. On the other hand, despite the thrill of harnessing the sun's power, the significance of taking safety precautions while installing solar panels must be addressed. This section digs into the significant part that safety plays in every stage of the solar installation process, beginning with the planning and site assessment stages and continuing through the component selection, installation, and ongoing maintenance stages.

Safety concerns are of the utmost importance during the design and site assessment phases of any solar project, which are the beginning stages of the project. To detect potential hazards and risks, it is vital to be completely aware of the energy requirements and the site's specific characteristics. When it comes to residential installations, it is essential to conduct a structural examination of the home. This serves the purpose of verifying that the roof or mounting structure can bear the weight of solar panels. Within commercial and industrial environments, the safety of individuals working in and around the installation area is an essential factor to consider. To reduce the risks associated with structural stability and electrical safety, site assessments should include crucial components such as shading studies, wind load estimates, and the proper grounding of equipment when necessary.

Regarding solar installation, selecting components is another crucial confluence where safety precautions are considered. When choosing solar panels, inverters, mounting structures, and energy storage options, they must comply with all applicable safety regulations and certifications. Equipment that has been certified reduces the likelihood of malfunctions, electrical failures, or fire hazards by ensuring that it complies with the rules that govern the industry. It is possible to quickly deactivate the solar power system when inverters are equipped with rapid shutdown characteristics, which is a safety feature that enhances safety during maintenance or emergencies. To construct a solar installation that is both secure and dependable, one of the most important steps is to make sure that all of the components meet or surpass the safety criteria.

In the beginning stages of the installation process, stringent safety measures are followed to ensure that solar panels, inverters, and other connected equipment are installed correctly. Technicians and solar specialists who are skilled in their field receive training to ensure that they adhere to the best practices in the industry and incorporate safety precautions into every stage of the installation process. To ensure the safety of installers working at elevated levels, it is usual practice to provide them with the appropriate personal protective equipment (PPE), including helmets, gloves, and safety harnesses. A reduction in the likelihood of accidents or falls occurring during rooftop installations can be achieved by implementing safe anchoring systems and observing occupational safety requirements.

Because electricity flows in both directions between the solar power system and the grid, grid-tied solar systems add an extra layer of safety considerations to the equation. Grid-tie inverters with anti-islanding characteristics prevent the system from electrifying the grid during power outages. This protects utility workers from the possibility of being electrocuted. Enabling the grid connection to be created securely, thereby limiting the risk of electrical dangers and enabling the seamless integration of solar power into the existing infrastructure, requires compliance with local electrical codes and coordination with utility providers.

Off-grid and hybrid systems, which feature energy storage options such as batteries, require special attention to safety during the installation process. There are unique safety considerations associated with chemical reactions, heat management, and ventilation associated with battery technology, which are often lithium-ion or lead-acid varieties. Installers ensure the safe installation and connection of batteries by adhering to the guidelines provided by the manufacturer and the standards established by the industry. The potential for thermal runaway or fire threats that are linked with battery storage can be reduced by the utilization of suitable enclosures, materials that are resistant to fire, and ventilation systems that are efficient enough.

The phase of the installation procedure known as commissioning and testing is essential, and it is at this period that safety precautions play a critical role. It is necessary to carry out performance testing, electrical safety checks, and functionality verifications to guarantee that the system functions within the parameters that have been defined. During the testing phase, installers ensure that lockout and tagout protocols are followed to prevent the system from being accidentally turned on. To ensure the overall safety and dependability of the solar power system, it is essential to conduct comprehensive inspections of the electrical connections, grounding systems, and protective devices.

Performing routine monitoring, maintenance, and inspections at regular intervals is essential to ensuring the continued safety of a solar project. Data on energy output and system performance can be obtained in real-time through monitoring systems, which can be integrated into inverters or used on separate platforms; whenever there is a variation from the predicted numbers or any indications of a malfunction, warnings are triggered, requiring rapid action. In routine inspections, which are typically carried out every year, a thorough examination of all components, wiring, and safety measures is carried out. Visual inspections, thermal imaging, and electrical tests all help in the early discovery of possible problems, which enables timely maintenance and repairs to be performed during the process.

The expertise and training of personnel who are involved in the process of installing solar panels is an essential component of safety, which goes beyond the physical components of the installation itself. Solar installers must undergo training programs and certifications emphasizing the significance of safety precautions, emergency response processes, and compliance with industry laws. Installers can maintain a safety culture within the solar industry by participating in ongoing education programs. These programs guarantee installers are current on the latest safety regulations, technology, and best practices.

In residential installations, where homeowners may play an active role in maintaining their solar systems, becoming aware of safety concerns becomes a responsibility shared by all parties involved. Homeowners receive training on safe measures, such as avoiding tampering with electrical components, performing routine visual inspections, and promptly reporting any possible problems. User manuals and ework that are easy to understand provide critical information on how the system operates, protocols for shutting down the system in an emergency, and contact information for expert assistance.

Emergency response planning is included in the safety factors considered in solar projects. Installers receive training to ensure that they can implement emergency shutdown procedures and respond to any threats, providing protection for themselves and others in the neighborhood. The readiness of installation teams to deal with unforeseen scenarios is further improved by the availability of emergency response equipment, such as fire extinguishers, and first aid training.

In conclusion, safety precautions are essential to every stage of the solar installation process. These precautions protect individuals, their particular property, and the community. A comprehensive approach to safety provides a safe transition to clean and sustainable energy by overseeing all aspects of the process, beginning with the planning and site assessment phase and continuing through the selection of components, installation, and continuous maintenance. As the solar business continues to grow, it is essential to prioritize safety concerns to cultivate public trust, reduce risks, and contribute to a robust and secure energy future. Not only does the commitment to safety protect individuals involved in solar installations, but it also highlights the industry's adherence to responsible practices in pursuing a more sustainable and linked world.

Troubleshooting Tips

The journey towards harnessing solar energy is an exciting and transformative endeavor, but like any technological system, solar installations may encounter challenges that require troubleshooting to maintain optimal performance. To recognize and resolve frequent problems in solar power systems, this section digs into the necessary troubleshooting strategies. To guarantee the longevity and effectiveness of solar installations, it is essential to have a thorough awareness of the complexities involved in troubleshooting. These complexities include performance disparities, electrical irregularities, environmental conditions, and equipment faults.

A reduction in the amount of energy produced or the system's performance is a problem that frequently arises with solar systems. Several causes can be attributed to this, with shading being a significant issue overall. The amount of energy solar panels produce is strongly impacted by shading, which might come from neighboring structures, vegetation, or other environmental obstructions. Identifying potential shading sources can be accomplished by conducting a shading study at the initial site inspection and then at regular intervals afterward. The strategic placement of solar panels or technology such as microinverters or optimizers can be utilized to maximize energy capture even in partially shaded settings. This can be done to mitigate shading difficulties.

It is also possible for solar panels to become contaminated with dirt and soil, which might compromise their performance. Dust, garbage, bird droppings, and other environmental contaminants can, over time, develop a coating on the panel's surface, reducing the amount of sunlight absorbed by the panel. Cleaning solar panels regularly, commonly done with water and a gentle brush or squeegee, helps ensure that they continue to maximize their performance. In areas that receive a low amount of rainfall and hence require cleaning less frequently, it may be beneficial to consider installing self-cleaning coatings or automated cleaning systems on panels to reduce the amount of soiling.

The electrical performance of solar arrays is another area frequently encountered as a source of worry. System owners or installers may observe voltage changes, energy generation anomalies, or grid connectivity problems. In circumstances like these, conducting a thorough electrical study is essential. Inspecting the connections between the wires, looking for damaged or loose cables, and ensuring the grounding is correct are all critical stages. Utilizing thermal imaging cameras allows for identifying possible hotspots in electrical components, which can indicate areas of increased resistance or places on the verge of failing. Electrical inspections should be performed regularly, particularly after severe weather events, as this helps identify potential problems earlier and prevents electrical hazards from occurring.

The malfunctioning of inverters is a particular area of troubleshooting frequently encountered in solar installations. The inverters, which are responsible for transforming the direct current (DC) electricity generated by solar panels into alternating current (AC) power that can be used, are an essential component of the system. The entire solar power system could be affected if an inverter fails. Real-time performance data is provided by monitoring systems that are integrated with inverters. This data enables the diagnosis of anomalies such as voltage variations, frequency deviations, or communication failures. When troubleshooting issues with an inverter, it is necessary to verify connections, update firmware, and, if required, follow the directions provided by the manufacturer or seek the assistance of a professional for repairs or replacements.

Suppose there is a breakdown in communication between components, particularly in systems connected to the grid. In that case, this can result in disturbances in the flow of energy in both directions. The process of troubleshooting communication problems may involve examining cables, making sure that the correct connection sequences are followed, and confirming that the communication protocols used by inverters and other components are compatible with one another. Inverters that are more advanced and equipped with error reporting features can provide insights into communication failures, which helps identify and resolve difficulties.

Additionally, environmental considerations provide difficulties in the installation of solar panels. Extreme temperatures, whether too high or too low, can affect the efficiency and lifespan of solar panels and the components linked with them. Due to the higher temperature coefficients, solar panels may face a decrease in efficiency in regions characterized by exceptionally high temperatures. Several passive cooling solutions can assist in decreasing heat-related performance losses. Some examples of these methods are raised mounting structures and well-ventilated installation designs. Removing snow from solar panels regularly is vital to restore energy output in regions that experience colder climates and where snow accumulation is an issue.

Troubleshooting issues associated with batteries are common in hybrid and off-grid solar installations, including energy storage devices. Issues such as capacity deterioration, voltage imbalances, and temperature instability are common problems with batteries, typically composed of lithium-ion or lead-acid molecules. To ensure the longevity and dependability of energy storage systems, it is vital to monitor the status of the charge, conduct capacity tests regularly, and appropriately perform temperature management. If a battery fails to function correctly, the problem can be resolved by speaking with the manufacturer, carrying out diagnostic tests, and, if necessary, replacing individual cells or the entire battery bank.

When it comes to troubleshooting, the function that monitoring systems and intelligent technologies play is an essential factor to consider. Intelligent monitoring platforms integrated with solar installations provide data in real-time regarding the production of energy, the performance of the system, and any potential problems. System owners or installers can rapidly resolve issues and undertake troubleshooting actions when they are provided with automated alerts and notifications. The preventive maintenance that these monitoring systems contribute to, which helps to minimize downtime and maximize the overall performance of solar installations, is a significant benefit.

When it comes to operations that include troubleshooting, the safety of personnel is of the utmost importance. It is of the utmost importance to adhere to safety protocols before conducting on-site inspections or interventions. These protocols include utilizing personal protective equipment (PPE) and deactivating the solar power system. To reduce the likelihood of electrical shocks or accidents occurring during maintenance, lockout and tagout procedures are implemented to ensure that electrical circuits are safely isolated. The rigorous safety training provided to maintenance professionals and installation highlights the importance of safe techniques in troubleshooting tasks.

In conclusion, troubleshooting in solar installations is a multi-step process that calls for a systematic approach to detecting and addressing problems influencing performance. Every circumstance requiring troubleshooting calls for a different set of diagnostic procedures and solutions. This includes anything from shading and soiling to electrical abnormalities and component problems. There is a strong correlation between the success of troubleshooting efforts and the implementation of preventative measures, comprehensive monitoring, and routine maintenance. As the solar industry continues to develop, refining troubleshooting procedures helps the durability and sustainability of solar installations. This ensures that solar installations will continue effectively generating clean and renewable energy for many years.

CHAPTER VII

Maintenance and Upkeep

Regular Inspections

Regular inspections are an essential component in the process of maintaining and optimizing the performance of solar installations. They play a crucial part in ensuring that these systems will last long and continue to function effectively. The proactive strategy of doing regular inspections is becoming increasingly important as solar energy continues to gain importance as an environmentally friendly and environmentally friendly alternative. This section investigates the diverse significance of routine inspections in solar installations. These inspections cover many factors, including system performance and safety, early problem detection, and the overall impact on the renewable energy landscape.

Regular inspections serve several important reasons, one of which is maintaining and improving the performance of solar installations. The efficiency of solar panels can be negatively impacted over time by several factors, including climatic conditions, soiling, shadowing, and wear and tear. Inspections at regular intervals offer a thorough picture of the system's health, which enables the identification and mitigation of faults that may influence energy production. This allows installers or owners of the system to implement preventative actions and ensure that the system functions at its highest possible level of performance. This is accomplished by evaluating the state of solar panels, inverters, and other associated components.

The amount of sunshine that solar panels are exposed to is directly proportional to the panels' efficiency. Shading, which might come from surrounding structures, grass, or other barriers, may considerably hinder the creation of more energy. During routine inspections, the installation site is subjected to a thorough investigation to identify potential sources of shading that may have established themselves over time due to changes in the environment surrounding the installation. This proactive strategy makes it possible to strategically prune vegetation, rearrange structures, or alter the solar array to maximize the amount of sunlight exposed to the variety and maximize the amount of energy captured.

Evaluating and correcting soiling on solar panels is another essential component of routine inspections that should be noticed. It is possible for a layer to build on the panel's surface due to the accumulation of dust, trash, bird droppings, and other environmental contaminants. This coating will reduce the quantity of sunlight that reaches the photographic cells. Cleaning at regular intervals, as specified by inspections, becomes necessary to keep efficiency at its highest possible level. Local environmental circumstances determine the frequency of cleaning, with arid or dusty locations requiring more frequent maintenance to ensure constant performance. The local ecological conditions determine cleaning frequency.

To guarantee the security and dependability of solar systems, it is essential to conduct routine inspections and consider performance factors. Because of the passage of time, electrical systems, inverters, wiring, and mounting structures are susceptible to wear, corrosion, and the possibility of damage. An in-depth evaluation of the electrical connections, the quality of the wiring, and the installation's overall structural stability are elements included in the inspection process. When loose connections, damaged cables, or signs of corrosion are identified, it is possible to make repairs promptly, reducing the likelihood of electrical faults or system failures.

When conducting inspections, grid-tied solar installations, which are solar panels connected to the electrical grid, require particular safety considerations. It is vital to perform routine tests on the grid connection and ensure compliance with local electrical codes to guarantee the safe incorporation of solar power into the existing infrastructure. Inverters equipped with safety features, such as the ability to shut down quickly, are subjected to inspections to ensure that they function correctly. To contribute to the overall reliability of solar installations and ensure that they conform with industry standards, installers and system owners should prioritize safety during predetermined checks.

Regular inspections provide several benefits, one of the most important of which is the early detection of problems, allowing for rapid action and mitigation. Integrating advanced monitoring systems into solar arrays can obtain real-time data on energy output, system performance, and potential abnormalities. During routine inspections, this data is subjected to a comprehensive analysis, which enables system owners or installers to spot deviations from the predicted values, variations in voltage, or anomalies in energy production. It is possible to reduce the amount of time that the system is offline and improve the overall reliability of the solar installation by resolving these issues as soon as they arise.

It is essential to pay particular attention to batteries during routine inspections because they are frequently necessary components in hybrid and off-grid solar installations. Battery banks significantly contribute to energy autonomy because they store excess energy when solar availability is low. To guarantee the efficient operation of energy storage systems, it is necessary to do routine checks on the capacity of the batteries, the voltage levels, and the charge-discharge cycles themselves. To undertake preventive maintenance and optimize battery performance, it is necessary to recognize potential problems early. These problems may include capacity decline, imbalances, or thermal instability.

The effects of environmental conditions are also a key contributor to the wear and tear that solar installations experience. Extreme weather conditions, such as storms, hail, or heavy snowfall, can make solar panels and mounting structures less reliable regarding their structural integrity. A comprehensive evaluation of any physical damage that may have been incurred due to unfavorable weather conditions is performed during routine inspections. It is possible to make prompt repairs or replacements to return the installation to its optimal condition. This will minimize the danger of additional deterioration and ensure the system will last for a long time.

In addition to the immediate benefits that are accrued by individual installations, conducting routine inspections contributes to the overall sustainability and efficiency of the renewable energy landscape. The whole can demonstrate the long-term viability of solar energy as a clean and reliable power source if solar installations are maintained in terms of efficiency and dependability. A significant contribution to the overall resilience of the renewable energy infrastructure is made by system owners, regardless of whether they are residential, commercial, or industrial. This is accomplished by engaging in frequent inspections and executing appropriate maintenance measures.

In addition, regulatory compliance is another aspect that is significantly impacted by the presence of regular inspections. Solar installations must adhere to local, regional, and national rules to comply with safety standards, environmental criteria, and grid interconnection norms. Regular inspections provide system owners and installers with a proactive tool to verify compliance, thereby lowering the risk of regulatory infractions and ensuring that solar installations are operated in a legal and ethical manner.

The solar business is continuously changing, and the incorporation of cutting-edge technologies into routine checks increases the efficiency of these inspections even further. Automated monitoring systems equipped with artificial intelligence and machine learning algorithms can analyze large amounts of data. This allows for the identification of patterns, trends, and

Difficulties that could arise. These technologies make it possible to perform predictive maintenance, which enables interventions to be based on data-driven insights. This reduces the need for reactive responses and further optimizes the efficiency of solar installations.

In conclusion, routine inspections are essential in the maintenance, performance optimization, and continued performance of solar installations. This approach's indispensability is shown by its several advantages, which include safety and performance concerns, early problem detection, and the capacity to ensure the continued viability of the sector. The commitment to regular inspections becomes a prudent approach for individual system owners and a collective responsibility to ensure the dependability and effectiveness of solar installations on a larger scale. Solar energy has firmly established itself as a central player in the global transition to sustainable power sources.

Cleaning Solar Panels

The use of photovoltaic systems to collect solar energy has emerged as a fundamental component in the ongoing effort to find renewable and environmentally friendly power sources worldwide. As the number of solar installations expanding on rooftops, in fields, and atop commercial structures continues to increase, it is becoming increasingly apparent that it is vital to maintain the efficiency of these installations. Even though cleaning solar panels may appear to be a simple chore, it has important implications for increasing energy output, assuring long-term performance, and contributing to the overall sustainability of solar energy. Within the scope of this section, the complex features of cleaning solar panels are investigated. The motivation for this practice, the issues it solves, and the approaches utilized to balance cleanliness and resource efficiency are all discussed.

The capacity of solar panels to accumulate sunlight is inextricably linked to the efficiency of these panels, which are responsible for converting sunlight into power. As time passes, solar panels can develop a layer of dust, dirt, bird droppings, pollen, and other contaminants found in the environment. The soiling on the panel surface acts as a barrier, reducing the amount of sunlight that reaches the photovoltaic cells and the amount of energy produced. Consequently, cleaning solar panels regularly becomes essential to ensure that they continue operating at their highest possible efficiency.

One of the most important aspects determining the frequency with which solar installations require cleaning is the placement of the installations geographically. It is common for the soiling rate to be higher in areas that are defined by dry climates characterized by dust and sand particles. As a result of the possibility that areas with minimal rainfall would see less natural cleaning, physical intervention is essential to guarantee constant functioning. On the other hand, areas that see a lot of rain might reap the benefits of natural cleansing, which could reduce the number of times that manual cleaning is required. Therefore, it is essential to have a solid understanding of the environmental variables present in a specific place to determine the suitable cleaning routine for solar panels successfully.

Even though it is indisputable that soiling affects the amount of energy produced, the rate at which dirt builds on solar panels might vary. The cleanliness of solar panels is affected by various factors, including the quality of the air in the surrounding area, factors such as the proximity to building sites and agricultural activities, and the presence of trees or vegetation. Pollution caused by industrial emissions and automobile traffic can contribute to the acceleration of soiling in urban environments. Similarly, solar installations located near agricultural areas may be susceptible to various types of particulate matter, such as pollen and dust from tilling. To maximize the efficiency of the cleaning process, it is essential to tailor the cleaning schedule to the particular environmental variables present at the installation site.

The frequency of cleaning solar panels is an essential factor to take into consideration since it requires finding a balance between the benefits of increased energy production and the avoidance of the consumption of resources that are not essential. According to the research findings, the appropriate cleaning frequency is contingent upon a wide range of criteria, such as the degree of soiling, the local climate conditions, and the cost of water. A study recently published in the journal Solar Energy highlights that in certain places, the energy gained from cleaning can offset the water consumption that is connected with it, making it a practical and environmentally responsible practice.

The overall environmental footprint of solar installations is impacted by the amount of water used and the amount of energy and resources utilized in the cleaning process. The long-term viability of solar panel washing is a topic of investigation in areas where there is a concern about the limitation of water resources. Innovative approaches, such as collecting rainwater or developing waterless cleaning technology, are being developed to reduce the negative impact on the environment connected with cleaning procedures that require a significant amount of water. To preserve the overall sustainability of solar installations, it is vital to find a balance between the benefits of cleaning and the resources consumed.

Solar panel cleaning encompasses various procedures, including operator manual cleaning, automated systems, and even self-cleaning technologies. These methodologies are used to clean solar panels. A standard method involves washing the surface by hand, usually accomplished with water and a gentle brush or squeegee. A hands-on evaluation of the cleanliness of each panel is made possible by this method, which is especially useful in areas with a modest amount of soiling. Nevertheless, it necessitates the utilization of labor, water resources, and continuous interventions.

The use of automated cleaning systems, designed to move around the solar array and apply water or cleaning solutions, provides a method that is more effective and does not require the use of hands. The operation of these systems, which are fitted with sensors and control mechanisms, can occur at predetermined intervals or in response to the conditions of the surrounding environment. However, even though automated cleaning reduces the amount of physical labor required, it does require an initial investment in the system's installation and upkeep. In addition, water utilization in many automated systems raises questions regarding water availability and the sustainability of water conservation.

Regarding solar panel maintenance, self-cleaning technologies offer a new sector undergoing development. These technologies use the intrinsic qualities of materials to repel dirt and prevent dirt from adhering to the panel surface. Examples of self-cleaning technologies include nano-coatings, hydrophobic treatments, and anti-reflective coatings. These technologies are designed to reduce the amount of dirt accumulating and make the cleaning procedure significantly more accessible. Even though these technologies have the potential to be widely adopted, their widespread adoption is dependent on several aspects, including their cost-effectiveness, their long- term durability, and their compatibility with various types of solar panels.

Several elements should be taken into consideration when selecting a cleaning method. These factors include the installation size, the soiling level, the environmental setting, and the water availability. Manual cleaning is better suited for smaller-scale residential installations, which allow homeowners to perform maintenance periodically. On the other hand, substantial utility-scale solar farms would reap the benefits of automated cleaning systems that can effectively cover vast areas with minimal involvement from manual labor. As these inventions evolve and demonstrate their effectiveness in real-world applications, adopting devices that generate their cleaning may become more widespread.

During the process of cleaning solar panels, safety is the most crucial factor to consider, in addition to concerns regarding efficiency and the utilization of resources. When it comes to accessibility, fall dangers, and the possibility of receiving an electrical shock, rooftop installations, in particular, present several issues. Professionals performing manual cleaning or maintenance tasks must adhere to safety protocols. These protocols include using personal protective equipment (PPE), secure anchoring systems, and compliance with industry safety regulations. The entire safety of the cleaning process is improved by incorporating safety elements into automated cleaning systems. These safety features include obstacle detection sensors and emergency shutdown functionalities, among other safety features.

To summarize, cleaning solar panels is an essential technique that plays a significant role in maximizing the amount of energy produced, ensuring that the panels will continue to function effectively over time and adding to the overall sustainability of solar installations. The decision-making process on the frequency of cleaning and the approach entails a complex review of environmental conditions, factors contributing to soiling, water availability, and the efficiency with which resources are utilized. During the solar industry's ongoing evolution, developing cutting-edge cleaning technology, environmentally responsible practices, and safety measures will play a crucial role in boosting the efficiency and lifetime of solar installations, further solidifying the industry's competitive position.

Replacing Components

The journey toward renewable and clean energy through solar installations is not just about the initial deployment of photovoltaic systems; it also involves a commitment to ongoing maintenance and the occasional repair of components. This is because solar installations are a renewable energy source. As solar technology continues to progress, replacing essential components to guarantee the lifespan, efficiency, and sustainability of solar installations during their lifetime is becoming increasingly important. This section dives into the many aspects of replacing components in solar installations. It investigates the logic behind replacement, the essential components that are prone to wear and degradation, the implications for the overall performance of solar energy systems, and their influence on the environment.

To comprehend the necessity of upgrading components in solar installations, it is necessary to acknowledge the dynamic and demanding environment in which these systems function. The solar panels, inverters, batteries, and other components involved with solar energy are subjected to various environmental conditions, including variations in temperature and humidity, ultraviolet radiation, and mechanical stress. These elements can, over time, contribute to wear, degradation, and the eventual malfunctioning of components, which will require their replacement to keep the solar power system's integrity and efficacy intact.

As the principal components responsible for energy generation, solar panels are constructed to resist exposure to outside elements. On the other hand, they are not beyond the influence of the effects of weathering and the circumstances of the environment. Solar panels gradually deteriorate with time, which is reflected in decreased efficiency and output, respectively. Even though contemporary solar panels are constructed to endure for several decades, their performance may drop. When the degradation reaches a point where the cost of lost energy output is more than the cost of replacement, the option to replace solar panels becomes strategic. Changing panels to improve their performance and energy yield is also possible. This can be accomplished by upgrading to panel technologies that are more efficient or by taking advantage of developments with photovoltaic materials.

There is a risk of wear and technological obsolescence for inverters, essential components for transforming direct current (DC) electricity generated by solar panels into alternating current (AC) power that may be used in houses or by the grid. The technologies used in inverters have undergone rapid development, resulting in increased energy efficiency, reliability, and functionality. The desire to take advantage of the advantages offered by more recent technology, to enhance the system's performance, or to address problems associated with the aging of current inverters may be the impetus for the decision to replace inverters. Since inverters are one of the most critical components in determining the total efficiency of a solar power system, replacing them at the appropriate time is essential to maximizing the amount of energy produced and optimizing the return on investment.

A fixed lifespan for batteries is governed by parameters such as charge-discharge cycles, depth of discharge, and operating temperatures. Batteries are often used in off-grid and hybrid solar setups to store excess energy for use during periods of poor solar output. When the batteries are getting close to the end of their life cycle, it is vital to replace them to keep the energy storage system reliable and to ensure that it can store resources. The development of new battery technologies may also lead to the introduction of replacements, providing an increased energy density, a longer cycle life, and improved safety features. Replacing batteries should consider the responsible disposal or recycling of old batteries as a critical aspect of reducing environmental impact.

Although they are not directly engaged in energy generation, mounting structures and racking systems are essential components that ensure the integrity and stability of solar installations. These components may experience corrosion, wear, or structural fatigue over time if they are subjected to the effects of environmental factors, temperature cycling, and mechanical forces over their lifetime. To guarantee the solar array's continuing security and stability, it is necessary to replace the mounting structures. Increasing the longevity of mounting structures and improving their functionality can be accomplished by upgrading to materials resistant to corrosion or incorporating enhanced tracking capabilities.

The establishment of the electrical circuits that link solar panels, inverters, and other components in solar systems is made possible by cables, connectors, and wiring, all essential components. Over time, these electrical components are subject to wear, corrosion, and damage since they are exposed to environmental elements and are susceptible to these things. Problems, such as weak connections, damaged cables, or evidence of wear, could be discovered by routine inspections and monitoring. Taking preventative measures such as replacing cables and connections is a preventive action that can be taken to resolve any electrical problems, limit energy losses, and ensure the safe and reliable operation of the solar power system.

When it comes to solar installations, making decisions regarding the replacement of components entails conducting an exhaustive analysis of several elements. It is essential to consider several factors when selecting when and which components should be changed. These factors include performance degradation, technical improvements, economic concerns, and system-specific requirements. The overriding objective is to find a middle ground between maximizing the effectiveness of the solar installation, controlling expenses, and contributing to the overall sustainability of solar energy.

The solar sector frequently experiences technological developments resulting in innovations in the design and functionality of parts. Newer generations of solar panels, inverters, and batteries offer increased efficiency, enhanced durability, and extra features that may not be included in previous components. These improvements are made possible by the availability of newer components. In particular, where the benefits of enhanced energy output or improved system reliability justify the cost, the decision to replace components may be driven by the desire to take advantage of these developments.

Economic factors are of the utmost importance when it comes to the decision-making process regarding the replacement of components. Considering the possible benefits in energy output and the overall return on investment, the decreasing costs of solar technologies have made replacement more financially feasible. This is especially true when considering energy production. By providing financial incentives or subsidy programs, the government may further encourage the replacement of older components with more modern technology that is more energy efficient.

This decision to replace components is also influenced by the requirements specific to the system and future plans. Suppose a solar installation is planned to undergo expansion or upgrading, for instance. In that case, it might be more feasible to replace specific components to guarantee compatibility and achieve the highest possible level of system performance. In addition, modifications to the patterns of energy use, the incorporation of energy-intensive appliances, or the incorporation of electric vehicle charging may require component updates to satisfy the ever-changing energy requirements.

When it comes to solar installations, the concept of replacing components is about enhancing performance and addressing environmental concerns. To reduce the adverse environmental effects, it is necessary to implement responsible recycling or disposal techniques when disposing of outdated components, notably batteries and electronic trash. The solar sector is placing a greater emphasis on developing sustainable and circular economy methods, with a particular focus on the recycling of materials and the reduction of electronic waste.

Another factor that should be considered is the amount of energy embodied in the process of producing and transporting new components. During the process of analyzing the environmental impact of component replacement, it is necessary to evaluate the energy and resource inputs that are necessary for the production, transportation, and installation of new components with the possible energy savings and efficiency benefits that could be achieved during the lifetime of the replacement component. One of the more delicate aspects of sustainable solar practices is balancing maximizing the environmental advantages of solar energy and minimizing the ecological footprint of component replacement.

Replacing components in solar installations is a strategic and vital feature of guaranteeing solar energy systems' sustainability, longevity, and efficiency. In conclusion, this makes the replacement of components an essential component. When deciding whether or not to replace components, it is necessary to conduct a thorough analysis of the performance deterioration, technological improvements, economic considerations, and environmental impact. This is because solar technology is constantly undergoing development. The solar sector can contribute to solar installations' sustained expansion and resilience by adopting responsible practices in component replacement. This will pave the way for a more sustainable and reliable future in terms of environmentally friendly energy.

CONCLUSION

In conclusion, "Solar Independence: A Comprehensive Guide to Off-Grid Solar Power - Harnessing Sustainable Energy for Off-Grid Living" is an invaluable resource for individuals seeking to embrace a sustainable and self-reliant lifestyle through off-grid solar power. This e-book provides knowledge from early design and site analysis to component selection, installation, and maintenance. It carefully navigates the complexity of off-grid solar systems.

The guide's title effectively sums up its contents, highlighting the technical details of off-grid solar power and the general goal of becoming independent from conventional power sources. By examining the particular difficulties and factors involved in living off the grid, the e-book equips readers to make appropriate decisions for their requirements and situations.

The promise of "Harnessing Sustainable Energy for Off-Grid Living" in the subtitle is kept throughout the guide by providing thorough insights into sustainable practices, environmental issues, and responsible resource management. The e-book gives readers a deeper grasp of the long-term advantages and ecological impact of installing an off-grid solar power system and provides them with the technical know-how to do so.

The guide's comprehensiveness guarantees that readers, regardless of experience level, may confidently begin their off-grid adventure. The e-book offers a step-by-step path, demystifying the complexity and providing helpful suggestions to handle problems, from system sizing to selecting the appropriate components.

In a world where self-reliance and sustainability are becoming increasingly important, "Solar Independence" shines as a lighthouse, pointing readers toward a way of life that embraces solar energy's potential while being environmentally conscious. As the e-book demystifies the nuances of off-grid solar power, it empowers individuals to adopt cleaner and renewable energy sources. It adds to the excellent discourse about sustainable living and the worldwide move towards a greener future. "Solar Independence" is a manifesto for those prepared to take the first steps toward achieving energy autonomy and a more sustainable way of living, not merely a set of guidelines.

Thank you for buying and reading/ listening to our book. If you found this book useful/ helpful please take a few minutes and leave a review on the platform where you purchased our book. Your feedback matters greatly to us.

Printed in the USA
CPSIA information can be obtained
at www.ICGtesting.com
LVHW021329110924
790643LV00014B/844

9 798869 147271